# Springer Theses

Recognizing Outstanding Ph.D. Research

## Aims and Scope

The series "Springer Theses" brings together a selection of the very best Ph.D. theses from around the world and across the physical sciences. Nominated and endorsed by two recognized specialists, each published volume has been selected for its scientific excellence and the high impact of its contents for the pertinent field of research. For greater accessibility to non-specialists, the published versions include an extended introduction, as well as a foreword by the student's supervisor explaining the special relevance of the work for the field. As a whole, the series will provide a valuable resource both for newcomers to the research fields described, and for other scientists seeking detailed background information on special questions. Finally, it provides an accredited documentation of the valuable contributions made by today's younger generation of scientists.

## Theses are accepted into the series by invited nomination only and must fulfill all of the following criteria

- They must be written in good English.
- The topic should fall within the confines of Chemistry, Physics, Earth Sciences, Engineering and related interdisciplinary fields such as Materials, Nanoscience, Chemical Engineering, Complex Systems and Biophysics.
- The work reported in the thesis must represent a significant scientific advance.
- If the thesis includes previously published material, permission to reproduce this must be gained from the respective copyright holder.
- They must have been examined and passed during the 12 months prior to nomination.
- Each thesis should include a foreword by the supervisor outlining the significance of its content.
- The theses should have a clearly defined structure including an introduction accessible to scientists not expert in that particular field.

More information about this series at http://www.springer.com/series/8790

Lene Kristian Bryngemark

# Search for New Phenomena in Dijet Angular Distributions at $\sqrt{s} = 8$ and 13 TeV

Doctoral Thesis accepted by
the Lund University, Lund, Sweden

 Springer

*Author*
Dr. Lene Kristian Bryngemark
Deutsches Elektronen-Synchrotron (DESY)
Hamburg
Germany

*Supervisor*
Prof. Torsten Åkesson
Lund University
Lund
Sweden

ISSN 2190-5053          ISSN 2190-5061   (electronic)
Springer Theses
ISBN 978-3-319-67345-5          ISBN 978-3-319-67346-2   (eBook)
DOI 10.1007/978-3-319-67346-2

Library of Congress Control Number: 2017952015

Printed on acid-free paper

This Springer imprint is published by Springer Nature
The registered company is Springer International Publishing AG
The registered company address is: Gewerbestrasse 11, 6330 Cham, Switzerland

# Supervisor's Foreword

The Large Hadron Collider (LHC) at CERN provides the energy frontier for Particle Physics research. It generates collisions between the proton constituents (quarks and gluons) that are studied using large detectors that measure the produced particles. The collision energies between the constituents have a statistical distribution since the sharing of the proton momenta between the constituents is also statistically distributed. Therefore, the more proton collisions that are sampled (estimated as integrated luminosity), the higher collision energies are reached.

This thesis is an analysis of the highest collision energies that the LHC produces, by studying the dijet angular distributions. This analysis is performed at both $\sqrt{s} = 8$ and 13 TeV proton-proton centre-of-mass energies.

Lene Bryngemark performed an excellent analysis demonstrating that up to the highest available momentum transfers between the colliding constituents, the Standard Model (QCD corrected to NLO-level with k-factors, plus electroweak corrections) perfectly describes the observations. She also showed very explicitly in the thesis how, in the data, the shapes of the angular distributions are unchanged over the full range of dijet invariant mass; the colliding objects appear the same across the range.

Having observed no deviation from the Standard Model, the analyses focused on using this information to set limits on a number of suggested models beyond the Standard Model. This was done for the compositeness scale, for the threshold mass of quantum black holes and for the mass of excited quark states. In addition, and for the first time, the angular distributions were used to set limits on the mass and coupling to fermions, of a dark matter mediator.

It is an exceptionally well-written thesis about a major analysis performed with the 8 TeV data and in particular the first 13 TeV data from the LHC at CERN. By performing a scattering experiment, the student has shown that the constituents remain pointlike up to the highest momentum transfers (smallest distances) available in a laboratory.

I don't know anyone easier and more pleasant to work with than Lene Bryngemark. She has shown drive, carefulness and a lot of independence. Lene also has shown strong leadership skills in the work with her colleagues, so I am sure she has a bright future in our field.

Lund, Sweden                                                      Prof. Torsten Åkesson
July 2017

# Abstract

A new energy regime has recently become accessible in collisions at the Large Hadron Collider at CERN. Abundant in hadron collisions, the two-jet final state explores the structure of the constituents of matter and the possible emergence of new forces of nature, in the largest momentum transfer collisions produced. The results from searches for phenomena beyond the Standard Model in the dijet angular distributions are presented. The data were collected with the ATLAS detector in proton-proton collisions at centre-of-mass energies of 8 and 13 TeV, corresponding to integrated luminosities of 17.3 $fb^{-1}$ and 3.6 $fb^{-1}$, respectively. No evidence for new phenomena was seen, and the strongest 95% confidence level lower limits to date were set on the scale of a range of suggested models. This work details the limits on the compositeness scale of quarks in a contact interaction scenario with two different modes of interference with Standard Model processes, as well as on the threshold mass of quantum black holes in a scenario with 6 extra spatial dimensions, and on the mass of excited quark states. It also includes new exclusion limits on the mass of a dark matter mediator and its coupling to fermions, as derived from the contact interaction limits using an effective field theory approach.

The performance in ATLAS of the jet-area-based method to correct jet measurements for the overlaid energy of additional proton-proton collisions is also presented. It removes the dependence of the jet transverse momentum on overlaid collision energy from both simultaneous interactions and those in the neighbouring bunch crossings and was adopted as part of the jet calibration chain in ATLAS.

## List of Publications

Some of the original work described in this thesis has appeared previously in the following publications:

[1] ATLAS Collaboration. "Search for New Phenomena in Dijet Mass and Angular Distributions from $pp$ Collisions at $\sqrt{s}$ = 13 TeV with the ATLAS Detector". In: *Physics Letters* B 754 (2016), pp. 302–322. ISSN: 0370-2693.

[2] ATLAS Collaboration. "Search for New Phenomena in the Dijet Angular Distributions in Proton-Proton Collisions at $\sqrt{s}$ = 8 TeV with the ATLAS Detector". In: *Phys. Rev. Lett.* 114 (2015), p. 221802.

[3] ATLAS Collaboration. "Performance of pile-up mitigation techniques for jets in $pp$ collisions at $\sqrt{s}$ = 8 TeV using the ATLAS detector". In: *Eur. Phys. J.* C76.11 (2016), p. 581.

[4] ATLAS Collaboration. "Search for new phenomena in the dijet mass distribution using $p$-$p$ collision data at $\sqrt{s}$ = 8 TeV with the ATLAS detector". In: *Phys. Rev.* D 91 (2015), p. 052007.

[5] ATLAS Collaboration. *Search for New Phenomena in Dijet Mass and Angular Distributions with the ATLAS Detector* at $\sqrt{s}$ = 13 TeV. Tech. rep. ATLAS-CONF-2015-042. Geneva: CERN, 2015.

[6] ATLAS Collaboration. *Pile-up subtraction and suppression for jets in ATLAS*. Tech. rep. ATLAS-CONF-2013-083. Geneva: CERN, 2013.

[7] ATLAS Collaboration. *Search for New Phenomena in the Dijet Mass Distribution updated using 13.0 fb$^{-1}$ of pp Collisions at $\sqrt{s}$= 8 TeV collected by the ATLAS Detector*. Tech. rep. ATLAS-CONF-2012-148. Geneva: CERN, 2012.

# Acknowledgements

It takes a village to write a thesis. Countless are the people who have contributed in one way or another to the work presented here, with efforts ranging from building detectors to cooking and caring for me when work has taken too much of my time. I will nevertheless attempt to list a few I want to thank.

Torsten, who suggested the research topic of dijet distributions and became my main supervisor. Always online, always there for physics or coffee or both, just waiting to make a discovery (I'm sorry we didn't yet). It is safe to say that without you, this thesis would not have been.

Else, Johan, my co-supervisors, who gave me the blessing of not worrying, but being encouraging and available when I needed it. A special thanks to them and David for thesis draft reading.

Ariel, who talks of jets with contagious enthusiasm and knows there is so much more we can do to extract all their secrets. I could not have had a better start in ATLAS than I got from working with you. I also thank John, who regrettably left physics, for patiently discussing and condensing all my (and Ariel's) wild pile-up ideas to a useful conclusion and strategy.

Caterina. Dijets, jets in ATLAS, just ATLAS would not be the same without you. Sharing code and expertise and cat pictures, you thoroughly showed me what working in a team can be. You're ready to both give advice and get your own hands dirty; Lund is better with you.

Tuva. You have to admit that this is one crazy adventure we started together. Our paths from starting a Bachelor's project to finishing a Ph.D. were increasingly parallel, but queerly similar—I can but don't want to imagine this time without you.

All my shorter and longer term colleagues (and you know I count students) who have made the division kitchen such a lively place! Anders, Bozena, Evert, Peter, the steady coffee and lunch crowd, always up for a debate and for support over the years. Florido and my fellow Ph.D. students (Alejandro, Anders, Anthony, Martin, Sasha, Tuva, Vytautas, and more recently, Ben, Edgar, Katja and Trine)—for coffee and beer, physics and politics, for lunches and dinners and the opportunity to vent our fair share of collaboration frustration. That same collaboration for making it all possible. The speedy dijet team, striking thrice within a year, ready to solve anything and on two occasions making our results public within weeks after data taking finished. The numerous delightful people in the university's most charming LGBTQ network for employees, you know who you are.

Hanno and Kroon and Laurie, for feeding and reading, for diverting my attention and listening to everything. Karin for laughter and all too rare long breakfasts and for always letting me leave the dishes for later. Pauline, for housing me during my year at CERN, for warm friendship and inviting me to join anything from dinners to running before dawn. Lotte for in turn housing my cat that year, and Julius-Lisa, that same cat for—I dare say—love and support.

The pursuit of a Ph.D. is the hardest on, and least gratifying for, the persons closest to you. Elina, words will never be enough. Thank you for letting me get absorbed and waiting for me to come out again. This one is for you.

# Contents

**Part I Introduction: The Thesis, The Standard Model, and The Experiment**

1    **Preamble** . . . . . . . . . . . . . . . . . . . . . . . . . . . . . . . . . . . . . . . . . . . . . .   3
    1.1    A Word on Particle Physics . . . . . . . . . . . . . . . . . . . . . . . . . . . . .   3
        1.1.1    Some Mention of the Scales . . . . . . . . . . . . . . . . . . . .   4
    1.2    The Energy Frontier . . . . . . . . . . . . . . . . . . . . . . . . . . . . . . . . . .   4
    1.3    This Thesis: Outline . . . . . . . . . . . . . . . . . . . . . . . . . . . . . . . . . .   6
    1.4    The Author's Contributions . . . . . . . . . . . . . . . . . . . . . . . . . . .   6
    References . . . . . . . . . . . . . . . . . . . . . . . . . . . . . . . . . . . . . . . . . . . . . . . .   7

2    **The Standard Model and Beyond** . . . . . . . . . . . . . . . . . . . . . . . . . . .   9
    2.1    Electromagnetism: QED . . . . . . . . . . . . . . . . . . . . . . . . . . . . . . .   11
        2.1.1    The Charged Leptons . . . . . . . . . . . . . . . . . . . . . . . . .   11
    2.2    The Weak (Nuclear) Interaction . . . . . . . . . . . . . . . . . . . . . . .   11
        2.2.1    The Neutral Leptons . . . . . . . . . . . . . . . . . . . . . . . . . .   12
    2.3    The Strong (Nuclear) Interaction: QCD . . . . . . . . . . . . . . . . .   13
        2.3.1    The Quarks . . . . . . . . . . . . . . . . . . . . . . . . . . . . . . . . .   14
    2.4    The Brout–Englert–Higgs Mechanism
        and the Particle Masses . . . . . . . . . . . . . . . . . . . . . . . . . . . . . .   14
    2.5    Antiparticles and Feynman Diagrams . . . . . . . . . . . . . . . . . . .   16
    2.6    Hadron Case Study: The Proton . . . . . . . . . . . . . . . . . . . . . . . .   17
        2.6.1    Parton Distribution Functions . . . . . . . . . . . . . . . . . .   18
        2.6.2    Perturbative QCD Calculations . . . . . . . . . . . . . . . . .   19
        2.6.3    Renormalisation . . . . . . . . . . . . . . . . . . . . . . . . . . . . .   20
        2.6.4    Factorisation Theorem . . . . . . . . . . . . . . . . . . . . . . . .   23
        2.6.5    Hadronisation . . . . . . . . . . . . . . . . . . . . . . . . . . . . . . .   23
        2.6.6    Underlying Event . . . . . . . . . . . . . . . . . . . . . . . . . . . .   24
    2.7    Monte Carlo Generators . . . . . . . . . . . . . . . . . . . . . . . . . . . . .   24
    2.8    Theories Beyond the Standard Model . . . . . . . . . . . . . . . . . . .   25
        2.8.1    Contact Interactions . . . . . . . . . . . . . . . . . . . . . . . . . .   26
        2.8.2    Quantum Black Holes . . . . . . . . . . . . . . . . . . . . . . . . .   27

|        | 2.8.3   | Dark Matter                                         | 28 |
|        | 2.8.4   | Excited Quarks                                      | 28 |
|        | References | ...                                             | 28 |

**3  The Large Hadron Collider** ........................................ 31
| 3.1 | Collider Kinematics | 32 |
|     | 3.1.1 Luminosity and Probability | 33 |
| 3.2 | Collider Data Taking | 34 |
| 3.3 | The LHC/beam Conditions | 34 |
| 3.4 | Pile-Up | 37 |
|     | References | 37 |

**4  The ATLAS Experiment** .............................................. 39
| 4.1 | Coordinate System | 39 |
| 4.2 | Collider Particle Detectors: The Onion Design | 41 |
| 4.3 | The ATLAS Detector Subsystems | 42 |
|     | 4.3.1 Magnets | 42 |
|     | 4.3.2 The Inner Tracker: Silicon Strips and Pixel Detector | 43 |
|     | 4.3.3 The Transition Radiation Tracker | 44 |
|     | 4.3.4 Calorimetry | 45 |
|     | 4.3.5 Muon Spectrometers | 45 |
|     | 4.3.6 LUCID | 46 |
|     | 4.3.7 More Forward: ALFA and ZDC | 47 |
| 4.4 | Detector Simulation | 47 |
| 4.5 | ATLAS Conditions | 47 |
|     | 4.5.1 Trigger System | 48 |
|     | 4.5.2 Data Quality | 48 |
|     | 4.5.3 Data Processing | 48 |
|     | References | 49 |

**Part II   Jets**

**5  Calorimetry** ........................................................ 53
| 5.1 | Electromagnetic Calorimetry | 53 |
|     | 5.1.1 Liquid-Argon Electromagnetic Calorimeter | 54 |
| 5.2 | Hadronic Calorimetry | 55 |
|     | 5.2.1 Tile | 56 |
|     | 5.2.2 LAr Forward Calorimeters | 57 |
| 5.3 | Resolution: Energy and Granularity | 57 |
| 5.4 | Energy Measurements | 59 |
| 5.5 | Noise | 62 |
| 5.6 | Topoclustering | 63 |
| 5.7 | Electromagnetic and Hadronic Scale | 63 |
|     | References | 64 |

**6 Jet Finding**........................................................ 65
  6.1 Sequential Recombination Algorithms ..................... 66
     6.1.1 $k_t$.......................................... 67
     6.1.2 Cambridge/Aachen ............................. 67
     6.1.3 Anti-$k_t$...................................... 67
  6.2 Jet Catchment Areas........................................ 67
     6.2.1 Active Area...................................... 68
     6.2.2 Passive Area: The Voronoi Area.................. 70
  6.3 Jets in ATLAS ............................................. 70
     6.3.1 Jet Calibration ................................. 72
     6.3.2 Jet Cleaning.................................... 73
  References......................................................... 74

**7 Pile-Up in Jets**.................................................... 75
  7.1 Pile-Up Observables........................................ 75
     7.1.1 Impact of Pile-Up on Jets ...................... 76
  7.2 Jet-Area Based Correction ................................. 77
     7.2.1 The $\rho$ Calculation: Algorithm Choices ............. 77
     7.2.2 The $\rho$ Calculation: $\eta$ Range..................... 78
  7.3 Method Performance....................................... 80
     7.3.1 Response....................................... 80
     7.3.2 Resolution ..................................... 89
     7.3.3 Jet Multiplicity ............................... 92
  7.4 Potential for Improvements................................. 94
  References......................................................... 95

**Part III Dijet Angular Distributions as a Probe of BSM Phenomena**

**8 Dijet Measurements**............................................... 99
  8.1 Dijet Observables ......................................... 99
     8.1.1 Dijet Kinematics .............................. 100
     8.1.2 Angular Distributions, $\chi$ ........................ 100
  8.2 Tools in the Analysis of Angular Distributions............. 102
     8.2.1 Comparing the Angular Distributions to Prediction.... 103
     8.2.2 Statistical Analysis ............................ 104
  8.3 Binning Considerations.................................... 106
     8.3.1 $\chi$ Binning ..................................... 106
     8.3.2 $m_{jj}$ Binning ................................. 106
  8.4 NLO QCD Corrections: $K$-factors........................ 106
  8.5 Dijet Mass Distribution................................... 109
  References......................................................... 110

**9    Signal Model Sample Generation** ............................ 111
    9.1    QCD ................................................ 111
    9.2    Contact Interactions ................................. 112
            9.2.1    $\Lambda$ Scaling ......................... 112
            9.2.2    Signal $K$-factors ........................ 112
            9.2.3    Normalisation ............................. 115
    9.3    Quantum Black Holes .................................. 116
    9.4    Excited Quarks ....................................... 117
    References ................................................. 117

**10   Analysis of Angular Distributions at $\sqrt{s} = 8$ and 13 TeV** ....... 119
    10.1    Event Selection ..................................... 119
            10.1.1    $\sqrt{s} = 13$ TeV ...................... 122
            10.1.2    $\sqrt{s} = 8$ TeV ....................... 122
    10.2    Corrections ......................................... 123
            10.2.1    Theoretical Corrections .................. 123
            10.2.2    Experimental Corrections: Removal of Masked
                      Modules .................................. 125
    10.3    Statistical Analysis ................................ 127
            10.3.1    Input .................................... 127
            10.3.2    Procedure ................................ 127
    10.4    Binning Optimisation ................................ 127
            10.4.1    $\sqrt{s} = 8$ TeV ....................... 128
            10.4.2    $\sqrt{s} = 13$ TeV ...................... 128
    10.5    Systematic Uncertainties ............................ 131
            10.5.1    JES ...................................... 131
            10.5.2    Luminosity Uncertainty ................... 133
            10.5.3    PDF Uncertainty .......................... 133
            10.5.4    Scale Uncertainty ........................ 134
            10.5.5    Tune Uncertainty ......................... 135
    10.6    Total Uncertainty ................................... 136
    References ................................................. 139

**11   Results** ................................................ 141
    11.1    Angular and Mass Distributions ...................... 141
            11.1.1    8 TeV .................................... 141
            11.1.2    13 TeV ................................... 145
    11.2    Statistical Analysis and Limits ..................... 148
            11.2.1    Fit Control Plots, Analysis of $\sqrt{s} = 13$ TeV Data ..... 148
            11.2.2    Limits on the Scale of New Phenomena ..... 149
    11.3    Discussion .......................................... 154
            11.3.1    Outlook for Methodology Improvements ..... 155
    References ................................................. 157

**12 Conclusions and Outlook** ................................. 159

**Appendix A: Simulation Settings** .............................. 161

**Appendix B: Data Set and Event Selection Details** ................. 165

**Appendix C: LHCP Results** .................................. 171

# Part I
# Introduction: The Thesis, The Standard Model, and The Experiment

# Chapter 1
# Preamble

–What is the smallest thing you know?

When asked what my research is about, I often find asking this question to be the most fruitful way to start. The answer varies, of course. Molecules, atoms, quarks? The smallest thing *I* know, is a mathematical point. This is a theoretical concept: just a point, a place-holder in some coordinate system, which is infinitesimally small—regardless of how much you zoom in, you will never see it; it has no extension in space. Mind-bogglingly, the particles I try to envision when doing my particle physics research are exactly this: point-like. They have mass, various charges, and they interact with each other, but they have no size. That is, to our current knowledge they don't. They are fundamental. And when you think about it, this is probably how it has to be: an entity with extension in space but still un-splittable, without constituents, is very difficult for the human mind to imagine. Conversely, a fundamental particle has no constituents, and thus no extension.

## 1.1 A Word on Particle Physics

Particle physics is the human endeavour to understand what the fundamental constituents of matter are, and how they interact. The programme is as simple as that. *Following* this programme is far from trivial: it takes building the largest instruments, fastest electronics, among the largest scientific collaborations and the coldest places in the Universe.[1]

At this point, all our knowledge and predictions about matter constituents and their interactions are neatly connected in the Standard Model of particle physics. Well, with one exception: this theory of the laws of nature does not include gravity. But it does include the three other interactions we have observed, and moreover, it does a

---

[1]Disclaimer: as far as we know—there could of course be some other civilisation somewhere achieving temperatures even closer to the absolute zero. But to be clear, we do know the temperature of outer space, and it is higher than what we use in some of our accelerators and experiments.

© Springer International Publishing AG 2017

L.K. Bryngemark, *Search for New Phenomena in Dijet Angular Distributions at √s = 8 and 13 TeV*, Springer Theses, DOI 10.1007/978-3-319-67346-2_1

splendid job describing them. The Standard Model will be described in greater detail shortly—suffice it to say here, that we know that it can still not be the final answer. This knowledge we base, quite simply, on the fact that we have more questions than it can answer. Some of the properties of the particles we observe—for instance, their masses—are not described in the Standard Model, but are free parameters that need to be experimentally established. Furthermore, there are several classes of observations indicating that there is a type of matter in the Universe which is not present in the Standard Model. Interestingly, this matter interacts with gravity, which is the only force of nature interacting with "normal" matter that is not included in the Standard Model.

### 1.1.1   Some Mention of the Scales

The matter around us is made up of atoms, which in turn consist of one or several electrons orbiting a nucleus made of one or several nucleons.[2] If we were to draw a simple picture of this system, what would be the relative scale of its pieces? If we draw the nuclear radius as 1 cm, then we would have to draw the electrons as infinitesimally small dots, about 1 km away. The quarks making up the proton don't seem to have a size yet either, but we know it's less than one thousandth of the proton's size—so on this sketch, it would be $10 \, \mu$m. Oh, and how large is the human scale on this drawing? 10 billion times larger than an atom—you would only need to draw a stack of 10 average European adults to cover the whole distance from the Earth to the Sun.

## 1.2   The Energy Frontier

With the start-up of CERN's new accelerator, the Large Hadron Collider (see Chap. 3), in 2009, a decade-long wait for the next energy leap was over. In the history of accelerators—which is the history of particle physics, since at least the 50s [1]—roughly speaking, when a new fancy accelerator was built, a new particle was found. This was true for instance for the Tevatron (the top quark) at almost 2 TeV and the SPS (the $W$ and $Z$ bosons) at 540 GeV. With the LHC, it took us three years to make our first discovery, after decades of planning: the $H$ boson. But we still hope for more.

At the basis of this relation (new accelerator = new particle), the most famous formula of physics—the Einsteins' $E = mc^2$—lies. In fact, it's not the new accelerator that is key. It's the new energy regime.

---

[2]*Nucleon*: nucleus constituent, that is, proton or neutron.

This formula is actually at the heart of our science. It states, that if we can produce enough energy, we can produce massive particles, since mass is a form of energy. In this game, mass is potential energy. Think of a rock held in your hand. Its potential energy with respect to gravity is released once you let go of it, and it falls to the ground, gaining kinetic energy as it falls. Similarly, a very massive particle often has potential energy with respect to another force field (recall the four fundamental forces of nature in our current description of nature) which is released as the particle transforms into lighter particles, generally with some kinetic energy—a decay.

Stop and think about it. We say that as we reach higher energies, we can produce heavier particles than ever before. This means, that the chance of finding something new, that was out of reach before, and which doesn't fit into our general picture (because our general picture worked fine as long as we didn't have to worry about this new thing) increases dramatically when we take a new energy leap. In one sense, we don't need to assume much: only that mass is a form of energy. But on the other hand, this new heavy particle must be able to communicate[3] with the incoming particles carrying this high energy. So in another sense, it's not a small requirement. Luckily, in quantum mechanics, generally all the things that are at all possible will happen eventually—it's just a matter of probabilities. And waiting.

Another aspect of being at the energy frontier is that higher energies correspond to resolving smaller details. This is another quantum mechanical feature: particles behave like waves, and waves like particles—it's a matter of at which energy scale you're looking. So, when we collide particles, the energy they have correspond to some wavelength. The higher the energy, the shorter the wavelength. And with a shorter wavelength, you can resolve smaller distances. Think of a boat lying in the sea: it will affect the pattern of the waves, which means that even if we wouldn't see the boat, we would be able to deduce that there was something in the water, some structure, from looking at the wave patterns. Now imagine a football floating next to the boat. This object is much smaller than the typical wavelength of the waves, and the wave pattern will not be distorted by its presence—we won't notice the ball. The same way, we can only resolve small details in the structure of matter if we have small enough waves, meaning, high enough energy. This means that for every leap in energy, we have a new possibility to resolve smaller structures in matter—effectively, to see if the particles we considered fundamental actually consist of something!

So what would you do with this knowledge? You know that the most probable things are already observed. You know that we have a new energy regime at our hands. You know that we can resolve smaller structures than ever before. And you know that in every collision, there is this, possibly small, quantum mechanical probability of any type of outcome allowed in nature. Well. I chose to study an enormous sample of the most energetically far-reaching type of outcomes: dijet events.

Here our journey begins. I set out to teach you all I know. I'm proud to say, it will take a little while.

---

[3]We assume a field, where information is carried by some mediator. If the mediator is recognisable by both sides, the transformation from kinetic energy to massive particle—and back to kinetic energy and lighter particles!—can happen.

## 1.3   This Thesis: Outline

The work presented here aims at using the collision final state of two *jets* (see
Part II) as a probe of phenomena beyond the Standard Model. The observable used
is the angular correlations of these two jets, an observable theoretically predictable
almost from first principles, and thoroughly studied at lower energies, including at
the LHC [2–8]. The road to such a measurement is however somewhat winding.
Here, with the privilege of retrospect, I will rearrange the dots so as to be able to
connect them with the shortest possible, continuous path, with the pattern finally
(and hopefully!) emerging clearly when I'm done.

In Chap. 2, the current best knowledge of particle physics, as described by the
Standard Model, is briefly outlined. Here the theoretical foundations needed for
the interpretation of the experimental results are laid. Then two chapters on the
experimental equipment: the accelerator (Chap. 3) and the detector (Chap. 4), follow.
We then switch gears and delve into the subject of measuring jets in Chaps. 5–7. The
last part of the thesis, Chaps. 8–11, comprises the description of the analysis method
details and the results from the dijet measurements made. Finally, the conclusions
follow in Chap. 12.

## 1.4   The Author's Contributions

The ATLAS experiment, which will be described later, is a large collaboration of
currently approximately 3000 physicists, and has been designed and constructed
for roughly two decades before it started producing papers about particle physics
measurements. All publications are made in the name of the collaboration. Hence,
only after a thorough internal review, the entire collaboration signs off on each
article, note, presentation and even poster made public. The author list, when shown,
is extensive, and follows strict alphabetical order.

Having a publication in your name is thus a slightly different game in this context
than in many other scientific communities. Firstly, one needs to qualify to become
a member of the author list. My qualification task[4] was to evaluate and, if useful,
introduce a new method to correct jet measurements for the impact of energy from
additional proton collisions (pile-up). This work will be detailed in Chap. 7, and
resulted in first a conference note [9], documenting the work in preparation for
presenting the results at a conference, and later in a paper [10]. I was one of two
main editors of the conference note, taking the initiative to start writing, and I wrote
the text describing the general concepts and ingredients of the method and the proof-
of-principle studies I made in simulation. Much of this text was later re-used for

---

[4]*Qualification task*: work done for the greater good of the collaboration, spanning at least 50% of
full working time over a year.

the paper, which also describes other aspects of improving jet measurements in the presence of pile-up. For all the assessment of the method in real data, I worked closely with the other authors. This method is now standardly used as part of the jet calibration chain in ATLAS, and thus underlies all ATLAS measurements involving (or vetoing on) jets using the 2012 data set or later. This illustrates the second aspect of the author list convention: every publication stands on the shoulders of countless hours of work by the (past and present) fellow members of the collaboration. Hence choosing a main author would be not only very difficult, but also very rude.

During 2012, I was part of the day-to-day detector operation, as *hardware on-call* for the Liquid Argon calorimeter (for more details on the calorimeters, see Chap. 5). I was on-call for approximately one quarter of the data taking over the year.

The next publication where I contributed directly to the measurement at hand was a conference note on the dijet mass resonance search[5] using part of the 2012 data set [11]. There I contributed the expertise I gained from the qualification task, in an investigation of the impact of pile-up on the measurement. I also contributed this knowledge to the full 2012 data set publication of the same search [12]. This measurement is closely connected to the dijet angular distribution search, where I was the main responsible for the search using 2012 data [13], and wrote the lion's share of the internal documentation used to assess the maturity of the analysis,and forming the basis for writing the paper.

The work done on the 2012 data set was a fantastic head start for doing two well prepared and very fast analyses [14, 15] of the first data coming out of the LHC in 2015, after its upgrade to higher energy. With my previous experience, I continued leading the analysis of the angular distributions, and took over most of the work preparing the theoretical predictions of the distributions (including the assessment of systematic uncertainties). I again wrote most of the internal documentation of these studies. This time the search was made in tandem with the mass distribution analysis, with joint leadership, strategy and documentation. I edited all parts of it, as well as the final paper.

# References

1. O. Chamberlain et al., Observation of antiprotons. Phys. Rev. **100**(3), 947–950 (1955)
2. UA1 Collaboration, G. Arnison et al., Angular distributions and structure functions from two jet events at the CERN SPS p anti-p collider. Phys. Lett. B **136**, 294 (1984)
3. UA2 Collaboration, P. Bagnaia et al., Measurement of jet production properties at the CERN collider. Phys. Lett. B **144**, 283–290 (1984)
4. D0 Collaboration, V.M. Abazov et al., Measurement of dijet angular distributions at $\sqrt{s} = 1.96$ TeV and searches for quark compositeness and extra spatial dimensions. Phys. Rev. Lett. **103**, 191803 (2009)
5. ATLAS Collaboration, Search for new particles in two-jet final states in 7 TeV proton-proton collisions with the ATLAS detector at the LHC. Phys. Rev. Lett. **105**, 161801 (2010)

---

[5]In contemporary particle physics nomenclature, a *search* is a measurement on data with the aim of discovering physics beyond the Standard Model.

6. CMS Collaboration, Search for quark compositeness with the dijet centrality ratio in 7 TeV pp collisions. Phys. Rev. Lett. **105**, 262001 (2010)

7. CMS Collaboration, Measurement of dijet angular distributions and search for quark compositiveness in pp collisions at $\sqrt{s}$ = 7 TeV. Phys. Rev. Lett. **106**, 201804 (2011)

8. ATLAS Collaboration, ATLAS search for new phenomena in dijet mass and angular distributions using pp collisions at $\sqrt{s}$ = 7 TeV. JHEP **1301**, 029 (2013)

9. ATLAS Collaboration, Pile-up subtraction and suppression for jets in ATLAS. Technical Report ATLAS-CONF-2013-083. Geneva: CERN (2013)

10. ATLAS Collaboration, Performance of pile-up mitigation techniques for jets in pp collisions at $\sqrt{s}$ = 8 TeV using the ATLAS detector. Eur. Phys. J. C **76**(11), 581 (2016)

11. ATLAS Collaboration, Search for new phenomena in the dijet mass distribution updated using 13.0 $fb^{-1}$ of pp collisions at $\sqrt{s}$ = 8 TeV collected by the ATLAS detector. Technical Report ATLAS-CONF-2012-148. Geneva: CERN (2012)

12. ATLAS Collaboration, Search for new phenomena in the dijet mass distribution using pp collision data at $\sqrt{s}$ = 8 TeV with the ATLAS detector. Phys. Rev. D **91**, 052007 (2015)

13. ATLAS Collaboration, Search for new phenomena in dijet angular distributions in proton-proton collisions at $\sqrt{s}$ = 8 TeV measured with the ATLAS detector. Phys. Rev. Lett. **114**, 221802 (2015)

14. ATLAS Collaboration, Search for new phenomena in dijet mass and angular distributions with the ATLAS detector at $\sqrt{s}$ = 13 TeV. Technical Report ATLASCONF-2015-042. Geneva: CERN (2015)

15. ATLAS Collaboration, Search for new phenomena in dijet mass and angular distributions from pp collisions at $\sqrt{s}$ = 13 TeV with the ATLAS detector. Phys. Lett. B **754**, 302–322 (2016). ISSN: 0370-2693

# Chapter 2
# The Standard Model and Beyond

It is often said that the Standard Model (SM) is a theory of interactions.[1] That means, that it describes the laws of nature by assigning its pieces a susceptibility to certain forces. This is modelled as a charge with respect to a field, which in this respect is nothing more than a quantum of how strongly it couples to the force carriers of that field.

The most familiar of charges is probably electric charge. Consider how static electricity separates the straws of your hair—this happens when there are a lot of same-sign charges repelling each other, a large total charge.[2] It does not happen when there are only local fluctuations up and down in charge, as there normally is (they largely cancel). The same way, the magnitude of the charge on a fundamental particle determines how strongly it is coupled to the corresponding field.

But how do the straws of your hair know about the electric charge of their neighbours? Well, the charge is communicated by the exchange of a messenger: a field quantum. The field quantum of electromagnetism is the photon—a particle of light. In every interaction in the SM, a field quantum is exchanged. These are commonly called gauge bosons. The different forces of nature in the SM all correspond to their own field, and are communicated with each their own set of gauge bosons. For gravity to fit into this picture, it too should probably be mediated by a particle: the stipulated *graviton*, which remains to be observed. In fact, that it is not observed, and that mass (the coupling to gravity) is not quantised, indicates that gravity cannot yet be described as a *quantum field theory* like the other forces of nature.[3] From now on, we will not consider gravity further, and as a matter of fact, we can safely neglect

---

[1] For a general introduction to the Standard Model, see for instance the review in [1], and references therein.

[2] A net charge arises as the hair is stripped of or receives *electrons*—fundamental particles with electric charge $-1e$. Unlike a compound object, a fundamental particle has an intrinsic, fixed charge.

[3] This could be an indication of a more fundamental theory than the SM.

© Springer International Publishing AG 2017
L.K. Bryngemark, *Search for New Phenomena in Dijet Angular Distributions at √s = 8 and 13 TeV*, Springer Theses, DOI 10.1007/978-3-319-67346-2_2

**Table 2.1** The four fundamental interactions currently known, their strength relative to the strong interaction at their respective appropriate scale, and range in metres [2]

| Force | Relative strength | Range (m) |
|---|---|---|
| Strong | 1 | $10^{-15}$ |
| Electromagnetic | $\frac{1}{137}$ | $\infty$ |
| Weak | $10^{-6}$ | $10^{-18}$ |
| Gravity | $10^{-39}$ | $\infty$ |

it, as it is many orders of magnitude weaker than the other three known forces of nature, which completely dominate particle interactions.

Moving from macroscopic compound objects like a straw of hair, the fundamental particles the SM deals with are *fermions* and *bosons*, with half-integer and integer (including zero) *spin*, respectively. Like charge, spin is a quantum number intrinsic to the particle, and it has a sign (is a directional quantity). In addition, a particle may carry charge under several fields, and thus interact with several forces. The combination of quantum numbers (spin type and charges) and mass[4] uniquely defines a fundamental particle. In total, the SM describes the interactions of 17 fundamental particles. The interactions and their range and relative strengths are listed in Table 2.1.

Although the table lists the properties of the fundamental interactions, let me immediately introduce a caveat. It so happens, that the strength of the interactions depends on the energy scale at which the interactions are probed. This is called "running of the coupling constants" and actually implies that at certain energies, forces can unite (unless they evolve exactly the same way). For instance at energy scales accessible to today's particle physics experiments, we often refer to electroweak[5] (EW) interactions.

As mentioned, the SM is a theory of interactions, and it is through the laws of interaction we can distinguish the particles. I will thus introduce the fundamental particles in the SM in terms of the interactions. It will become evident that some interactions and prediction techniques are more relevant to my work, as they will be described in greater detail, and will serve as a use-case for some of the general features of the SM formalism. Mathematically, the SM is also a theory of symmetries; from symmetries, interactions and conservation laws arise. Conservation laws have profound implications on the interpretation of the theory, but are also part of our experimental tool-box, as they allow us to deduce certain quantities that aren't directly observed.

---

[4]Here it is again, the elusive, seemingly fundamental, concept of mass.

[5]*Electroweak* as in the unification of electromagnetic and weak interactions.

## 2.1  Electromagnetism: QED

Magnetism has been known by humanity for thousands of years, and even used (e.g. for navigation). Electricity was understood as a force much later, in the 19th century. The electron would be the first particle discovered which is still considered fundamental.

In the quantum world, electromagnetism is described by Quantum ElectroDynamics (QED). Its mediating gauge boson is the photon (often represented by a $\gamma$ (gamma)). It is an infinite-range force, since the mediator is mass- and chargeless. This is the force which keeps atoms together, from the opposite electric charge sign of electrons and atomic nuclei. It also governs the electromagnetic waves we encounter in our everyday lives in form of radio (cell phone) signals, visible light or X-rays.

QED is one of the most tested theories we have—that is, we can both predict and measure quantities very precisely. The energy in an atomic energy level transition in hydrogen is often quoted as an example, as it is measured to 14 digits [3]! Yet, as we shall see, it is not a complete theory to all scales.

### 2.1.1  The Charged Leptons

Here we encounter our first matter particle type: the electrically charged leptons. One of these, the lightest, is the aforementioned electron ($e$). It partly makes up matter as we know it in our everyday life. However, it has heavier cousins: the muon, $\mu$, and the tau lepton, $\tau$. These cousins have different *flavour*, and different mass, but apart from that they are similar. Flavour is a quantum number that is conserved under the electromagnetic interaction. The charged leptons have unit electric charge.[6]

## 2.2  The Weak (Nuclear) Interaction

The weak interaction is suitably named, as it is substantially weaker than both the strong and electromagnetic interaction. It is mediated via massive vector bosons, the electrically charged $W$ and the neutral $Z$ boson, and unlike electromagnetism, it can transform particles into a cousin of different flavour. The masses of the gauge bosons make it a short range force. The weak interaction charge is called *weak isospin*,[7] and it is only carried by particles of *left-handed chirality*.

---

[6]The electron being the first fundamental particle discovered, it set the standard for electric charge—as the name suggests.

[7]In the unified electroweak force, the charge is instead *weak hypercharge*, which takes both weak isospin and electric charge into account.

A particle of right-handed helicity is one where spin orientation and direction of motion coincides, while for a left-handed particle these two are opposite. This means that handedness depends on the reference frame of the observer. For massless particles, there is no choice of two frames with respect to which the massless particle can appear to move in opposite directions, since no observer can travel faster than the particle. Thus they are always of definite helicity, which coincides with its chirality. For massive particles, only chirality is invariant of choice of reference frame. This "handedness" or chirality is necessary to explain certain experimental observations, such as parity violation.[8]

### 2.2.1  The Neutral Leptons

Along with the weak interaction, the need for neutral leptons—*neutrinos*—arises. They are ordered in flavour doublets[9] together with the charged leptons as illustrated below, in order of increasing mass:

$$\begin{pmatrix} e \\ \nu_e \end{pmatrix}_L \quad \begin{pmatrix} \mu \\ \nu_\mu \end{pmatrix}_L \quad \begin{pmatrix} \tau \\ \nu_\tau \end{pmatrix}_L$$

As for the neutrino masses themselves, they are too small to have been directly measured yet. That neutrinos do have mass is however established through the phenomenon of neutrino oscillations: neutrinos produced in one flavour state can oscillate into another flavour state[10] as they travel. And travel they do! Since they only carry charge under the weak interaction, they rarely interact, and are very likely to travel straight through even large macroscopic objects like planets.

The weak interaction can convert an upper particle in a doublet to its lower counterpart. This is possible since there are charged weak bosons, $W^\pm$, which can carry the incoming charge such that it is overall conserved. For instance, in radioactive $\beta$ decay, it is the weak interaction which is at play: $n \to p + e^- + \bar{\nu}_e$ involves the exchange of a $W$ boson. But to understand that process, we first need to introduce a set of particles commonly associated with the last known fundamental force of nature.

---

[8]We won't need to discuss parity further in this work, but for a historical experiment, the interested reader is referred to Ref. [4].

[9]$L$ denotes left-handed. The right-handed counterparts are flavour singlets, and thus stand alone: $e_R, \mu_R, \ldots$.

[10]Flavour oscillations are a quantum mechanical subtlety, relating to the flavour *eigenstate* not being the same as the mass eigenstate. Oh, yes, there it is again.

## 2.3 The Strong (Nuclear) Interaction: QCD

In our everyday lives, the main effect of the strong interaction is to keep the atomic nuclei together. This is not a small impact! The strong interaction is however a short-range force, limited to within the size of a nucleon, and only a smaller residual force is actually felt between the nucleons.

*Colour charge* is the quantum number making particles susceptible to the strong interaction or colour force, described by Quantum ChromoDynamics (QCD). The colour charges are, in an analogy to the components of white light, *red, green* and *blue*, expressed below in a colour triplet:

$$\psi_a = \begin{pmatrix} \psi_1 \\ \psi_2 \\ \psi_3 \end{pmatrix} \tag{2.1}$$

The gauge boson of the strong interaction is the *gluon*. Gluons carry colour charge themselves. Thus, in contrast to QED, where the photon does not carry electric charge, two gluons can interact. This in turn makes the range of the strong interaction finite even though gluons are massless.

The QCD Lagrangian, the equation of motion describing all of the workings of the theory, is formulated in a gauge invariant way as

$$\mathcal{L} = \mathcal{L}_q + \mathcal{L}_g = \bar{\psi}_a (i\gamma^\mu \partial_\mu \delta_{ab} - g_s \gamma^\mu t^C_{ab} A^C_\mu - m\delta_{ab})\psi_b - \frac{1}{4} F^{\mu\nu}_A F^A_{\mu\nu}, \tag{2.2}$$

where Eq. 2.1 enters, and the field tensor

$$F^A_{\mu\nu} = \partial_\mu A^A_\nu - \partial_\nu A^A_\mu + g_s f^{ABC} A^B_\mu A^C_\nu \tag{2.3}$$

makes up the kinetic term in the gauge field. The third term of Eq. 2.2 makes $\bar{\psi} i \not{\partial} \psi$ gauge invariant. Gauge invariance is a means for making local symmetries in a theory evident, and in practice it means that a given new choice of coordinate system must be accompanied by a choice of *covariant* derivatives (the $\partial_\mu$ for instance), such that there is no net change on the predictions of the theory. The physics doesn't change! But the choice of formalism can make it more or less obscure. Since local symmetries give rise to forces, this is a central point in the Lagrangian formulation. On a similar note, global symmetries correspond to conserved currents, or put more simply, conservation laws.

In Eqs. 2.2 and 2.3, the eight[11] gluons enter in the $A^1_\mu, \ldots, A^8_\mu$, accompanied by the eight generators $t_{ab}$ and the structure constants $f^{ABC}$. The superscripts here are colour indices implicitly summed over. From the strong coupling strength, $g_s$, we

---

[11] $8 = 3^2 - 1$, QCD being an $SU(3)$ symmetry group.

define the strong coupling constant $\alpha_s = g_s^2/(4\pi)$. The last term in Eq. 2.3 is the self-interaction term due to the colour charge of the gluons.

### 2.3.1   The Quarks

The six quarks are fermions—building blocks of larger compounds of particles. They carry colour charge, meaning they belong in colour triplets, and non-integer[12] electric charge: *up* ($u$), *charm* ($c$), *top* ($t$) have $+\frac{2}{3}e$, while *down* ($d$), *strange* ($s$) and *bottom* ($b$) carry $-\frac{1}{3}e$. Note that gluons carry one colour and one anti-colour, giving them the possibility to change the colour state of for instance a quark in an interaction. None of the other fermions in the SM interact via the strong interaction—they are colourless, or colour singlets. Like the leptons, the quarks also come in three generations, ordered in flavour doublets as represented below, again ordering the doublets in increasing mass:

$$\begin{pmatrix} u \\ d \end{pmatrix}_L \quad \begin{pmatrix} c \\ s \end{pmatrix}_L \quad \begin{pmatrix} t \\ b \end{pmatrix}_L$$

From this structure, it should be clear that the quarks also carry weak isospin and take part in weak interactions. However, due to the much smaller weak interaction coupling strength, QCD processes are much more probable and thus happen more often.

## 2.4   The Brout–Englert–Higgs Mechanism and the Particle Masses

No thesis covering work done in ATLAS in recent years would be complete without mentioning the Brout–Englert–Higgs (BEH) mechanism, and the related $H$ boson discovered by ATLAS and CMS in 2012. This mechanism gives masses to the fermions and weak gauge bosons via the mechanism of electroweak symmetry breaking, splitting the massless gauge bosons of the underlying symmetry into the massless photon and the massive $W$ and $Z$ bosons, thus splitting the electroweak theory into electromagnetic and weak interaction. Knowing at which energy we have unification, we could predict approximately what the mass of the $H$ boson should be, even though mass is always a free parameter in the SM.

In the general picture of quantised coupling strengths, the $H$ boson is a little special since the coupling to different particles is related to their mass. Or, conversely, the mass of a particle is a measure of—given by!—how strongly it couples to the BEH

---

[12]Had the history of discovery been different, the electric charge of the electron had likely been defined as $-3e$ instead.

**Table 2.2** The masses of fundamental particles as experimentally measured, or in most quark cases, calculated [2]. Note that the light quark masses are current quark masses, as calculated in the $\overline{\text{MS}}$ scheme at a scale of 2 GeV

| | Particle | Symbol | Mass |
|---|---|---|---|
| Leptons | Neutrinos | $\nu_e, \nu_\mu, \nu_\tau$ | < 25 eV |
| | Electron | $e$ | 511 keV |
| | Muon | $\mu$ | 105.6 MeV |
| | Tau lepton | $\tau$ | $1776.2 \pm 0.1$ MeV |
| Quarks | Up | $u$ | $2.3^{+0.7}_{-0.5}$ MeV |
| | Down | $d$ | $4.8^{+0.5}_{-0.3}$ MeV |
| | Strange | $s$ | $95 \pm 5$ MeV |
| | Charm | $c$ | $1.275 \pm 0.025$ GeV |
| | Bottom | $b$ | $4.18 \pm 0.03$ GeV |
| | Top | $t$ | $173.21 \pm 0.51 \pm 0.71$ GeV |
| Bosons | Photon | $\gamma$ | 0 |
| | Gluon | $g$ | 0 |
| | Charged weak | $W$ | 80.4 GeV |
| | Neutral weak | $Z$ | 91.2 GeV |
| | Higgs boson | $H$ | $125.7 \pm 0.4$ GeV |

field. In relativity, mass governs how fast[13] something can travel at a given energy. Nothing travels faster than light in vacuum, precisely because photons are massless. And even though the BEH field permeates even the vacuum, photons don't interact with it and remain massless. Other particles can't travel as fast, as they are interrupted by having to interact with the medium. It is actually very similar to light in an atomic medium, such as glass. Here light travels more slowly than in vacuum, which gives glass its refractive index. At an atomic level, what happens is that the photon is constantly absorbed and re-emitted, slowing it down. On top of that, it is emitted in any random direction. From quantum mechanical effects, however, the sum of all possible paths introduces a lot of cancellations, and one direction of a light ray will be the final one. The final effect is that the light ray has refracted. In the process, the photons were moving more slowly, which can be thought of as acquiring an effective mass. Analogously, particles interacting with the BEH field acquire their masses too—the only difference being, that this medium exists **everywhere**. The masses of the fundamental particles as currently known are listed in Table 2.2.

For comparison, the proton and neutron weigh in at about 1 GeV. It is obvious that there are many fundamental particles which are heavier than these composite ones! Why the masses differ by up to five orders of magnitude between the fundamental particles is indeed a mystery in the present theoretical system.

---

[13]The relation between energy and velocity is given by $E^2 = m^2 + \vec{p}^2$.

**Fig. 2.1** Feynman diagram illustrating $e^- e^- \rightarrow e^- e^-$ scattering, under the exchange of a photon ($\gamma$)

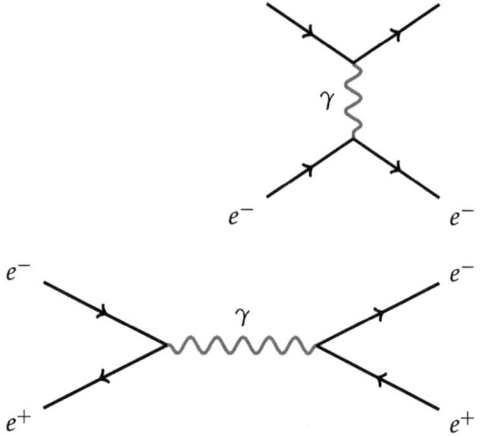

**Fig. 2.2** Feynman diagram illustrating $e^+ - e^-$ annihilation into a photon ($\gamma$), and pair production back into an $e^+ - e^-$ pair

## 2.5 Antiparticles and Feynman Diagrams

For all of the fermions, there are also antiparticles, with the opposite sign on charges (charge conjugation). These are, for the electrically charged leptons, simply denoted with a $+$ instead of a $-$: the electron $e^-$ has an anti-particle $e^+$. For neutrinos and quarks, antiparticles are denoted with a bar: $u$ and $\bar{u}$.

The seemingly simple concept of antiparticles is still a crucial ingredient in charge conservation: only if the net charge is equal before and after the interaction, a transformation from energy in the form of one set of particles to another can occur. This is achieved in the annihilation or creation of particle-antiparticle pairs, where the net charge is 0 both before and after the interaction.

To guide intuition, there is the useful construct of a Feynman diagram. It has a profound interpretation in terms of probabilities of different processes, but let's focus on its illustrative strengths for now. In these diagrams, time flows from left to right, lines represent particles, and each vertex represents an interaction. Fermions are represented with solid straight lines, with arrows pointing right for particles and left[14] for antiparticles. Gauge bosons are represented with wavy or curly lines for electroweak bosons and gluons, respectively. Figure 2.1 is our first encounter: it illustrates how two electrons interact with (repel) each other under the exchange of a photon, the gauge boson of QED. As mentioned before, this gauge boson exchange is the model for how particles are affected by each other's presence.

Figure 2.1 shows a "space like" process. If we rotate the diagram by 90°, we get a "time like" process, as shown in Fig. 2.2.

Guided by the direction of the arrows, we realise that what is depicted in Fig. 2.2 is particle-antiparticle annihilation and pair production. The mass energy of the par-

---

[14]This convention goes back to considering antiparticles as particles moving backwards in time, as introduced in [5].

ticles is converted into photon energy. This is in turn converted back into a particle-antiparticle pair. As long as the available energy is large enough, a vertex like this can go in any direction (creation as well as annihilation). There is no requirement that the photon conserves flavour; it has no memory thereof as its flavour quantum number is zero (as is the combined positive and negative flavour quantum numbers of the electron and anti-electron[15]). As long as the other vertex conserves the flavour content, by for instance creating a muon-antimuon pair which taken together has zero flavour, all is well, and if the energy of the photon is large enough to create the mass of two muons, this can happen.

## 2.6  Hadron Case Study: The Proton

At this point, we have covered all the fundamental particles. But there is one more particle that is important to consider here: the proton, which we use for particle collisions. The proton is one example of a hadron[16]—a particle composed of quarks. Being composite, it is a suitable strong interaction case study, and we will use it to introduce some additional concepts. This is however a fairly complex topic, and we need to split it into pieces.

While quarks carry colour, hadrons as a whole are colourless. This can be accomplished in two ways: by a combination of colour-anticolour (e.g. a red-antired) as in *mesons*, or in a combination of all three (anti)colours red–green–blue, as in *baryons*. Hadrons thus consist of two or three (anti)quarks.[17] These are called valence quarks. In addition, there always occur quantum fluctuations[18] where a gluon splits into

---

[15] Anti-electron: also known as *positron*.

[16] The concept of hadrons is older than the quark model, so, they must have certain unique characteristics, evident already before.

[17] Colourless combinations thereof, such as *pentaquarks*, have also been observed [6].

[18] Virtual particles can "borrow" additional energy from the vacuum, but only for a short time.

**Fig. 2.3** Feynman diagram
illustrating what nuclear $\beta$
decay looks like at quark
level, if one could resolve the
$W$ boson

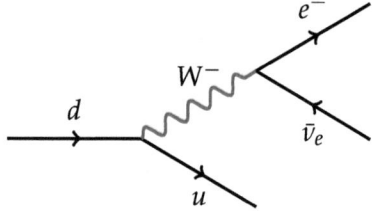

a quark-antiquark pair which then annihilate back into a gluon. These fluctuation quarks are virtual, or *sea*, quarks.

Firstly, we establish that the proton is a baryon: it consists of three valence quarks, *uud*. This gives the proton a net electrical charge of $+1e$, and as mentioned before, no net colour charge. The other baryon making up ordinary matter, the neutron, has valence quarks *udd*, making it electrically neutral. The neutron is slightly heavier than the proton,[19] and an isolated neutron thus decays to a proton. At quark level, the transformation from $d$ to $u$ would imply weak decay involving a $W$ boson, as illustrated in Fig. 2.3. There is in general not enough energy to create real $W$ bosons when this happens, only virtual or *off-shell* $W$ bosons that immediately produce a real lepton and neutrino. The comparatively long life-time of the isolated neutron, $\sim$13 min, reflects all of this.

### 2.6.1   Parton Distribution Functions

Since the proton is a composite particle, if we accelerate the proton to carry a certain momentum, it is its constituents that carry this net momentum. The motion of constituents inside the proton is not restricted and can be both lateral and longitudinal, but the net effect has to be the overall proton momentum. We can thus stipulate

$$\sum_i \int x q_i(x) dx = 1, \qquad (2.4)$$

where the $x$ is the Bjorken $x$ [7], which is the longitudinal momentum fraction carried by a parton, and the sum is over the quark indices $i$. We have already touched upon the concept of sea quarks, originating from quantum fluctuations inside the protons. By denoting proton as *uud*, we mean that we get a non-vanishing result

$$\int (u(x) - \bar{u}(x)) \, dx = 2 \qquad (2.5)$$

and

---

[19]More strictly speaking: $m_n > m_p + m_e + m_{\bar{\nu}_e}$.

$$\int \left( d(x) - \bar{d}(x) \right) dx = 1 \tag{2.6}$$

when we integrate over all the $q$ and $\bar{q}$ content of the proton. The number of accessible sea quark flavours depends on the energy scale at which the proton is probed. This immediately means that the fraction of the proton momentum carried by gluons and sea quarks, respectively, depends on the energy transfer $Q$ in the collision that probes the proton structure. In fact the fractions vary also for the valence quarks. Overall, the quarks and the gluons carry about half the momentum each. The fractions are given in the Parton Distribution Function (PDF). Two examples at different $Q^2$ are shown in Fig. 2.4, which shows that when the proton is probed at larger momentum transfer, the valence quarks become increasingly less dominant also at higher $x$. Albeit not theoretically known per se, the PDF evolution with $Q^2$ can be calculated from a given starting point using the Dokshitzer-Gribov-Lipatov-Altarelli-Parisi (DGLAP) equations. The starting point has to be an experimental measurement of the PDF at some $Q^2$. This can be data from for instance electron-proton or proton-proton collisions since the proton structure itself is universal and not dependent on the type of experiment. However, in the former case only one proton PDF is probed, making the extraction of information a little less involved.

### 2.6.2 Perturbative QCD Calculations

The logic of the Feynman diagrams, with a vertex for each interaction and mediating particles, easily lends itself to *perturbation theory*. Perturbative calculations split

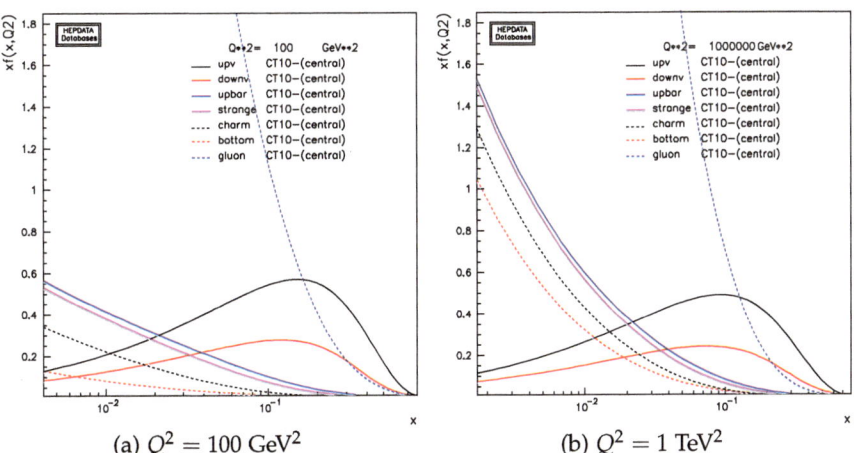

(a) $Q^2 = 100$ GeV$^2$          (b) $Q^2 = 1$ TeV$^2$

**Fig. 2.4** PDFs using NLO predictions including LHC data, for two values of $Q^2$: **a** 100 GeV$^2$ and **b** 1 TeV$^2$ [8, 9]

complicated calculations in pieces of increasing fine tuning, and start with the coarsest approximation. The method is to make an expansion[20] in increasing orders of your variable in a region where it is small, such that higher order contributions rapidly get smaller. In practice, a suitably truncated expansion is often good enough—luckily, since higher-order corrections are often not known, or computationally expensive, for a complicated expression.

Considering a process illustrated by a Feynman diagram, there is generally more than one way to draw it; there is more than one imaginable way to go from a given initial to final state, with more or less complicated steps in between. In quantum mechanics, we can't distinguish different possible histories—the intermediate steps in a process—leading up to a measured final state. But they all happen, with some probability! In a full calculation of the probability of an outcome, all of these possible paths need to be calculated, and summed correctly taking quantum mechanical interference into account. But in a Feynman diagram every vertex represents an interaction with a coupling strength, and all the vertices are multiplied to give the total probability, or *cross section*. This means that two different paths, with a different total number of vertices, are at different orders in coupling strength. If the coupling strength is small enough—which, as we shall see shortly, is the case for the small-distance, high energy transfer collisions explored in this thesis—the more complicated paths contribute increasingly little to the final result. In a perturbative calculation of the cross section of the process, we can thus truncate the expansion at some level of complexity without much loss of precision! Perturbation theory holds already for $Q > 1$ GeV, which is the proton mass and approximate confinement scale in QCD. Often the leading, or lowest, order (LO) result is a good approximation, but the next-to-leading order (NLO)corrections can be substantial.

## 2.6.3  Renormalisation

As mentioned, when applying the Feynman rules, all possibilities have to be integrated over, and they often come with momenta in the denominator. This gives rise to divergent (infinite) integrals, which would have to be cut off at some finite scale $\Lambda$ to give finite results. Mathematically, this is not isolated to quantum field theories, even if it is a common feature of them.[21] Rather, it arises when one makes an expan-

---

[20]The idea is similar to the method of Taylor expansion.

[21]This discussion loosely follows Ref. [10], which gives an overview of the renormalisation idea that is worth a read!

sion of a dimensionless quantity (e.g., a probability) around a small dimensionless parameter (say, coupling strength) of a function that depends on a dimensional parameter (for instance momenta). To remain dimensionless, the calculated quantity has to depend on the dimensional parameter through the ratio with another parameter of the same dimension—a *regulator*, say, $\Lambda$. After choosing a regularisation scheme, one can redefine couplings, masses and other parameters to absorb the divergences. Typically the redefinition corresponds to a physically measured quantity (such as a coupling constant) at a given scale, which we call the renormalisation scale $\mu_R$, with the dimensions of mass. In practice what happens is that the implicit dependence on $\Lambda$ in the original expansion was removed. Only after this, we let $\Lambda \to \infty$ and get finite results. The price paid in this procedure is that the coefficients in the perturbative expansion only make sense in a given context of scale and corresponding coupling. In addition, we must abandon thinking of parameters as constant: when a quantity normalised at one scale is measured at a very different scale, the couplings and masses adjust. Also, the $\Lambda$ introduced as an upper cut-off of the integrals to remove the divergence, can be thought of as the scale at which the physical theory no longer holds—a scale at which *new physics* enters.[22]

**The Running of $\alpha_s$**

This immediately brings us to the question of the strong coupling constant. As indicated above, its value will depend on the scale at which we measure it. Experimentally, the value of $\alpha_s$ is given at the Z mass, and the world average is $\alpha_s(M_Z) = 0.1185(6)$ [2]. The scale dependence of $\alpha_s$ is controlled by the $\beta$ function, which is precisely one of those parameters which do not depend on $\Lambda$:

$$\alpha^2 \frac{d\alpha_s}{d\alpha^2} = \beta(\alpha_s) = -(b_0\alpha_s^2 + b_1\alpha_s^3 + \mathcal{O}(\alpha_s^4)), \tag{2.7}$$

where $b_0 = (33 - 2n_f)/(12\pi)$, $b_1 = (153 - 19n_f)/(24\pi^2)$, and $n_f$ is the number of accessible quark flavours. If we let $\alpha^2 = Q^2$, we can express the effective coupling strength as $\alpha_s(Q^2)$, where $Q$ is the scale of the momentum transfer in the process at hand. Equation 2.7 shows a negative evolution of the coupling constant with the renormalisation scale $\mu_R$. The implications are even more evident in the expression for $\alpha_s$ itself: from the $\beta$ function, we obtain

$$\alpha_s(Q^2) = \frac{4\pi}{b_0 \ln(Q^2/\Lambda_{QCD}^2)} \cdot$$
$$\cdot \left[ 1 - \frac{2b_1}{b_0^2} \frac{\ln[\ln(Q^2/\Lambda_{QCD}^2)]}{\ln(Q^2/\Lambda_{QCD}^2)} + \mathcal{O}\left( \frac{1}{\ln^2(Q^2/\Lambda_{QCD}^2)} \right) \right] \tag{2.8}$$

---

[22]For QED the physically meaningful upper cut-off is the scale of unification with the weak interaction.

**Fig. 2.5** Schematic illustration of the factorisable processes in a *pp* collision, where one parton from each proton undergoes a hard scattering

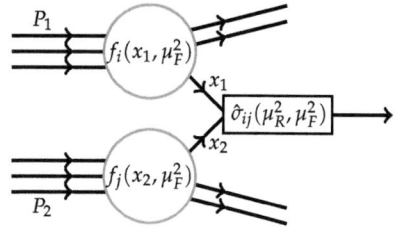

Here the reference scale $\Lambda_{QCD} \sim 200$ MeV is the confinement scale of QCD: this is the limit where $\alpha_s$ diverges and becomes strong. In this regime, the perturbative approach is no longer valid. In the limit $Q \to \infty, \alpha_s \to 0$. In between these regimes, $\alpha_s$ depends only logarithmically on $Q$. Furthermore, it is immediately clear that also the $\alpha_s$ value will depend on the order to which the perturbative expansion is carried out.

### Confinement and Asymptotic Freedom

A possible way to think of a physical cause of the running coupling constant is in terms of (anti-)screening. Consider an electron. Just like a gluon fluctuates in and out of sea quark pairs, an electron constantly emits and reabsorbs field quanta, most likely photons. This can in turn create virtual loops of electron/positron pairs, which screen the charge seen farther from the electron. The net effect is a smaller effective charge of the electron, making the field around it weaker. Similarly, gluons are constantly emitted from and reabsorbed by the quark. These can in turn create virtual gluon loops, which enhance the field strength at a distance, but smear the quark colour charge as we look closely. So, the strong interaction coupling "constant" depends on the distance, or equivalently energy,[23] at which it is probed. At smaller distances (higher energies) $\alpha_s$ is smaller. In fact, at higher energies, more pair production becomes possible—this is one way of seeing why the classical (or leading order) approach breaks down: as we need to consider more possible paths, we need to introduce renormalisation.

The small coupling constant at high energies is called asymptotic freedom: at small distances, well inside the hadron, partons barely interact and are very loosely bound. As two quarks are increasingly separated, the potential binding energy increases. In fact the potential between them increases linearly—much like in a classical spring or rubber band, a picture exploited in the Lund string model [11], which we will summarise shortly. This theoretically requires a non-Abelian term, causing self-interactions.[24] Confinement means, that one can never observe a free quark.

---

[23] In the *natural units* commonly used in particle physics, where the speed of light in vacuum $c = 1$, distance has dimensions of 1/(energy).

[24] The electroweak theory is also non-Abelian, and W and Z bosons are self-interacting. Photons are not.

### 2.6.4 Factorisation Theorem

We concluded that we can use perturbative calculations for the high-energy processes that we are generally interested in. We have also seen, that the effective energy at which we are probing the proton, and as a result the rate of the process, depends on the PDFs. These are however not possible to calculate perturbatively, which mathematically manifests itself as divergent integrals. But luckily, the two regimes are independent—they are *factorisable*. This means that we can rely on the calculation of the DGLAP evolution for the non-perturbative PDF part, and do perturbative calculations of the hard scatter part, without loss of generality. Technically this introduces a *factorisation scale* $\mu_F$, with $1 \text{ GeV}^2 \leq \mu_F^2 < Q^2$. For the regime below the factorisation scale, we use the non-perturbative proton quark distribution. The hard-scatter cross section $\hat{\sigma}_{i,j}$ is governed by short-distance processes and perturbatively calculable. We can then express the cross section for a hard scatter in a hadronic collision factorised as

$$\sigma(P_1, P_2) = \sum_{i,j} \int dx_1 dx_2 f_i(x_1, \mu_F^2) f_j(x_2, \mu_F^2) \hat{\sigma}_{i,j}(\alpha^2, \mu_F^2), \qquad (2.9)$$

where the $P_{1,2}$ denote the incoming hadron momenta and the participating partons carry $p_1 = x_1 P_1$, $p_2 = x_2 P_2$. The $f_{i,j}(x, \mu_F^2)$ are the PDFs at some given Bjorken $x$, as given at the factorisation scale. This factorisation is schematically illustrated in Fig. 2.5.

### 2.6.5 Hadronisation

Since only colourless particles can travel macroscopic[25] distances, an outgoing parton from a hard scatter has to *hadronise*. This is a non-perturbative process, occurring at lower energy and correspondingly larger distances than the hard scatter, where $\alpha_s$ is large.

In the Lund string model, the force between two partons is pictured as a string. It has the properties of a classical string in the sense that the field contains a constant

---

[25]Macroscopic—or even outside the proton radius.

amount of field energy[26] per unit length, meaning that the potential increases linearly when the string is stretched [11]. If two quarks are pulled apart, in for instance a high energy collision, the binding energy becomes so large that it is energetically "cheaper" to create a real quark-antiquark pair between them, which breaks the string without resulting in free quarks (but in new strings between quarks and anti-quarks). This process is repeated as long as there is sufficient energy. The end result is a collimated hadron shower, called a *jet*, in the direction of the original quark. This jet essentially carries the energy, momentum and other properties of the original quark. Note that since hadronisation happens at longer time scales than the hard scatter process, it can't affect the partonic cross section of a process, or violate conservation laws. Measuring the jet properties is thus the way to access the properties of the original quark, even if it can't be isolated and measured itself. It is also a good way to measure their interactions.

### 2.6.6   Underlying Event

The remaining piece of our proton case study, is the remnants of the proton itself after a hard scatter involving one of its partons. In a violent high-energy collision, an outgoing parton produces jets due to confinement, as we have seen. Similarly, the proton remnants (illustrated in Fig. 2.5) acquire colour in the collision, and will undergo similar hadronisation. The remnants, however, often travel along the direction of the incident proton, and predominantly produce soft and diffuse radiation as measured in the transverse direction to the beam.

## 2.7   Monte Carlo Generators

In order to discern deviations from the expected SM behaviour in the processes studied, we need to make predictions of the SM. Our theoretical framework allows for perturbative calculations to finite orders, and non-perturbative processes such as hadronisation will remain. Using a Monte Carlo (MC) event generator, we can obtain a (pseudo-)random representation of the possible outcomes in for instance a proton collision, mimicking the stochastic processes by sampling a probability distribution. Complete generators will model both the hard-scatter process and parton showers (initial and final state radiation), hadronisation, multiple interactions and underlying event, providing a list of produced particles and their four-vectors at a given stage of the process. There are also incomplete generators calculating the hard-scatter cross sections only, which in turn may provide these calculations to higher orders.

---

[26]The colour field lines are not radial (as in electromagnetism) but compressed in a flux tube between the partons.

**Fig. 2.6** Feynman diagram illustrating the unresolved interaction leading to $q\bar{q}$ production in a Contact Interaction (CI) approach

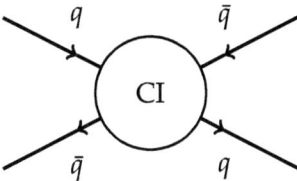

The underlying hypotheses for the non-perturbative processes giving these distributions can vary: the widely used complete MC generator PYTHIA [12] uses the Lund string model. This is the main generator used for the work described in this thesis.

## 2.8 Theories Beyond the Standard Model

There are numerous proposed extensions of the SM, intended to answer one or more of the outstanding questions posed by observations that seemingly have no fundamental explanation in the existing theoretical framework. Particle masses are, as I may have hinted before, a free parameter in the SM which still seems to be of some profound importance, especially if we want to unify all the known forces of nature. There are also numerous independent observations of phenomena that tell us that only about 5% of the total energy content in the universe is matter as we know it, and as all theories used in any field of science describe it. There is evidence that there is about five times as much *Dark Matter* as normal matter; the rest of the energy content in the universe is considered to be *Dark Energy* [13], the general properties of which are completely unknown. Finally, there is no a priori knowledge that the particles considered fundamental right now would not in fact have constituents—the history of particle physics actually points in the other direction. One could also argue that the mass hierarchy and generational structure points to fermion compositeness. All in all, the SM seems to be an effective theory holding up very well at the scales and the precision at which we have been probing it so far, but it may eventually have to yield to a more complete description of nature.

The measurements described in this thesis would be sensitive to many of the new phenomena predicted by such proposed extended theories. The strategy relies on simple yet powerful assumptions on what we can expect from SM processes, and the primary goal is to quantify the deviations in data from the SM prediction, rather than discover a specific hypothesised new phenomenon. Here I will focus on describing the so-called *benchmark models* used in the analysis: models making distinct predictions of observable distributions compared to the SM. When comparing these predictions to measured data, we can often make statements about the degree of compatibility with data, given certain parameter values in the model. Thus we learn something even from not discovering anything new: we learn how we *can't* describe nature.

**Fig. 2.7** Feynman diagram
illustrating the effective field
theory approach to $\beta$ decay

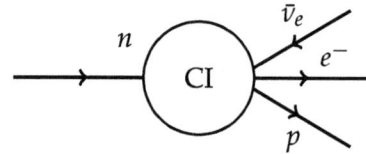

### 2.8.1  Contact Interactions

One way to model fermion compositeness is to consider that at some energy scale, a
new force of nature becomes manifest, as we resolve what is keeping the composite
particles together. Well before that energy, however, there may be an effect on the
probability and kinematic characteristics of a process, such as jet production. We
can thus discern that there is something new before resolving the details of the
process. This situation can be satisfactorily modelled with an *Effective Field Theory*
(EFT), as depicted in Fig. 2.6. Actually, this approach is the same as in the four-point
interaction of Fermi, describing nuclear $\beta$ decay when there is not enough energy
to resolve the W boson exchange. This is drawn in Fig. 2.7. In such a description, a
scale $\Lambda$ is introduced, dictating at which point we can resolve the processes hidden
in the circle—the Contact Interaction (CI) [14–16]. It follows that as $\Lambda$ grows, the
signal strength gets weaker, if we keep the probe energy constant. The description
chosen in this work is an additional effective Lagrangian:

$$\mathcal{L}_{CI}(\Lambda) = \frac{g^2}{2\Lambda^2}[\eta_{LL}\left(\bar{q}_{iL}\gamma^{\mu}q_{iL}\right)\left(\bar{q}_{jL}\gamma_{\mu}q_{jL}\right)$$
$$+ \eta_{RL}\left(\bar{q}_{iR}\gamma^{\mu}q_{iR}\right)\left(\bar{q}_{jL}\gamma_{\mu}q_{jL}\right) \tag{2.10}$$
$$+ \eta_{RR}\left(\bar{q}_{iR}\gamma^{\mu}q_{iR}\right)\left(\bar{q}_{jR}\gamma_{\mu}q_{jR}\right)],$$

where $i(j)$ is a flavour index, $g$ denotes the strong coupling strength, and $\eta = 0, \pm 1$
represents the sign of the interference between CI and two-quark initial and final
states of QCD: + for destructive and—for constructive interference.[27] The CI is
characterised by the compositeness scale $\Lambda$ and its mode of interference with the
QCD $q\bar{q} \rightarrow q\bar{q}$ process, where constructive interference is overall expected to lead
to an enhanced signal, while for destructive interference, the effects of signal and
interference compete. The CI modelling leads to non-resonant enhancement (or suppression) of jet production.

---

[27]Sign convention; confusing but true.

## 2.8.2 Quantum Black Holes

In a scenario where gravity propagates in more dimensions than the other fundamental forces, it would be diluted, causing it to appear much weaker than the others [17, 18]. This mechanism thus provides an explanation to the experimental observation that gravity is weaker than the other forces. The full set of space-time dimensions is commonly referred to as the *bulk*, while particles interacting under the SM[28] live on the *brane*, a 4D hypersurface in the $4 + n$ dimensional scenario. The number $n$ of extra dimensions vary between realisations; typically $n = 1$ in a Randall–Sundrum (RS) scenario [17] and $n = 6$ in an Arkani-Hamed–Dimopoulos–Dvali (ADD) scenario [18]. This would in both cases lower the fundamental scale of gravity, $M_D$,[29] from the *Planck scale* $M_{Pl} \sim 10^{18}$ GeV to the vicinity of the electroweak scale $m_{EW} \sim 100$ GeV, which is clearly accessible at the LHC (see Chap. 3). This idea elegantly solves the so-called *hierarchy problem* in the SM, which is the question why these two seemingly fundamental scales are so widely separated, and it does so without introducing any new symmetries or interactions but by instead changing the space–time metric.

It does however introduce the possibility that microscopic or Quantum Black Holes (QBHs) are produced at the LHC. A TeV scale black hole created in a collision would decay to bulk and brane particles, giving an experimental possibility to detect it. If the black hole mass is larger than $M_D$, the black hole will thermalise and decay to high-multiplicity final states; however, there are many reasons to suspect that this is not the first mode of discovery, but rather 2-body final states are, as suggested in Ref. [19] and briefly summarised here.

Firstly, since they have not been discovered yet, it is unlikely that the energy threshold needed has been surpassed in previous experiments. Secondly, there is large suppression of Bjorken $x$ through the PDFs, and energy loss from the initial parton-parton system, pushing the available black hole masses down for a given available centre-of-mass energy. In a regime below the production threshold energy, strong gravitational effects enhance the 2-body final state cross section through exchange of a mediating particle produced in strong gravity, even if the final state is not a black hole. Finally, even for cases with larger multiplicities, it may be seen as contrived to assume a complexity in which not also 2-body final states would be enhanced. Even so, a multi-body final state would still contribute to an analysis of 2-body final states which doesn't impose an upper limit on the number of final state objects.

---

[28]Note that since they don't interact in the SM, right-handed neutrinos are here not constrained to stay on the brane!

[29]The naming conventions and parameter choices vary between models. Here we choose the ADD representation.

### 2.8.3 Dark Matter

There is very little known about the properties of Dark Matter (DM). It interacts gravitationally, and makes up about 1/4 of the energy content in the universe—a factor 5 more than the normal matter (at least partly) described by the SM. DM particles remain to be detected.

A common approach in collider searches for DM is to assume that the DM particles produced in a collision escape detection. However, for them to be produced in the first place, there has to be a production mechanism involving coupling to partons, leaving a non-zero probability also of jet production. The production mechanism is often modelled in an EFT approach, where the scale of the phenomena is too high to be resolved using the available collision energy. This resembles the treatment of CI outlined above. Care must however be taken to avoid using an EFT approach in the regime where the available energy is larger than the scale of the new phenomenon. Here a simplified theory, assuming a mediator with some mass and a set of coupling strengths to fermions and dark matter, is a more suitable approach.

### 2.8.4 Excited Quarks

One consequence of quark compositeness would be the possibility of excited quark ($q^*$) states. Deexcitation proceeds through the emission of a gluon, making a resonant $qg$ final state, since excitation energies would be discrete. Excited quark production and subsequent decay to quarks and gluons via gauge interactions has been used as a common benchmark for the dijet mass resonance search [20–24], and it is described in detail in Refs. [25, 26]. It is used in this thesis as a representative model for resonant dijet production.

### References

1. G. Altarelli, The standard model of particle physics. CERN-PHTH-2005-206 (2005)
2. K.A. Olive et al., Particle data group. Chin. Phys. C **38**, 090001 (2014)
3. M. Niering et al., Measurement of the hydrogen 1S–2S transition frequency by phase coherent comparison with a microwave cesium fountain clock. Phys. Rev. Lett. **84**(24), 5496–5499 (2000)
4. C.S. Wu et al., Experimental test of parity conservation in beta decay. Phys. Rev. **105**(4), 1413–1415 (1957)
5. E.C.G. Stueckelberg, Remarks on the creation of pairs of particles in the theory of relativity. Helv. Phys. Acta **14**, 588–594 (1941)
6. LHCb Collaboration, Observation of $J/\psi p$ resonances consistent with pentaquark states in $\Lambda_b^0 \rightarrow J/\psi K^- p$ decays. Phys. Rev. Lett. **115**, 072001 (2015)
7. J.D. Bjorken, Asymptotic sum rules at infinite momentum. Phys. Rev. **179**, 1547–1553 (1969)
8. H.L. Lai et al., New parton distributions for collider physics. Phys. Rev. D **82**, 074024 (2010)
9. The Durham HepData Project. PDF Plotter (2016), http://hepdata.cedar.ac.uk/pdf/pdf3.html

10. B. Delamotte, A hint of renormalization. Am. J. Phys. **72**, 170–184 (2004)
11. B. Andersson et al., Parton fragmentation and string dynamics. Phys. Rep. **97**(2–3), 31–145 (1983). ISSN: 0370-1573
12. T. Sjostrand, S. Mrenna, P.Z. Skands, A brief introduction to PYTHIA 8.1. Comput. Phys. Commun. **178**, 852 (2008)
13. Planck Collaboration, Planck 2013 results. I. Overview of products and scientific results. Astron. Astrophys. **571**, A1 (2014)
14. E. Eichten et al., Supercollider physics. Rev. Mod. Phys. **56**, 579 (1984)
15. E. Eichten et al., Erratum: supercollider physics. Rev. Mod. Phys. **58**, 1065 (1986)
16. P. Chiappetta, M. Perrottet, Possible bounds on compositeness from inclusive one jet production in large hadron colliders. Phys. Lett. B **253**, 489 (1991)
17. L. Randall, R. Sundrum, A large mass hierarchy from a small extra dimension. Phys. Rev. Lett. **83**, 3370–3373 (1999)
18. N. Arkani-Hamed, S. Dimopoulos, G. Dvali, The hierarchy problem and new dimensions at a millimeter. Phys. Lett. B **429**(3–4), 263–272 (1998). ISSN: 0370-2693
19. P. Meade, L. Randall, Black holes and quantum gravity at the LHC. JHEP **05**, 003 (2008)
20. ATLAS Collaboration, Search for new particles in two-jet final states in 7 TeV proton-proton collisions with the ATLAS detector at the LHC. Phys. Rev. Lett. **105**, 161801 (2010)
21. ATLAS Collaboration, Search for quark contact interactions in dijet angular distributions in pp collisions at $\sqrt{s} = 7$ TeV measured with the ATLAS detector. Phys. Lett. B **694**, 327 (2011)
22. ATLAS Collaboration, Search for new physics in dijet mass and angular distributions in pp collisions at $\sqrt{s} = 7$ TeV measured with the ATLAS detector. New J. Phys. **13**, 053044 (2011)
23. ATLAS Collaboration, Search for new physics in the dijet mass distribution using 1 fb$^{-1}$ of pp collision data at $\sqrt{s} = 7$ TeV collected by the ATLAS detector. Phys. Lett. B **708**, 37–54 (2012)
24. ATLAS Collaboration, Search for new phenomena in the dijet mass distribution using pp collision data at $\sqrt{s} = 8$ TeV with the ATLAS detector. Phys. Rev. D **91**, 052007 (2015)
25. U. Baur, I. Hinchliffe, D. Zeppenfeld, Excited quark production at hadron colliders. Int. J. Mod. Phys. A **2**, 1285 (1987)
26. U. Baur, M. Spira, P.M. Zerwas, Excited quark and lepton production at hadron colliders. Phys. Rev. D **42**, 815–825 (1990)

# Chapter 3
# The Large Hadron Collider

The Large Hadron Collider (LHC) [1] is a 27 km circumference storage ring with counter-rotating bunched proton or lead ion beams.[1] It is located 100 m below ground at CERN outside Geneva, Switzerland. Some of the design specifications are given in Table 3.1.

The proton acceleration is staged in several steps, starting from the hydrogen source and over pre-acceleration using previous generations of CERN accelerators. The LHC is thus the last collider in a larger accelerator complex. When the protons are injected in the LHC the beam energy is already 450 GeV. From there the protons are further accelerated and the accelerator optics focus and defocus the bunches to optimise for efficient collisions and long beam lifetime. Once stable beams are declared, the lateral beam spot width in the ATLAS experiment, introduced below, is of $\mathcal{O}(10)$ μm.

Similarly, reaching the design goals mentioned in Table 3.1 is done in incremental steps. During 2012, the beam energy was 4 TeV, with a bunch spacing of 50 ns. After Run1 a 2-year upgrade shutdown followed, and Run2 began when the LHC started delivering beam at 6.5 TeV each, with a bunch spacing of 25 ns,[2] in May 2015. As seen in the table, there are many empty proton bunch spaces—or *Bunch Crossing Indices*, BCIDs—foreseen. The 25 ns bunch spacing refers to the time spacing between BCIDs. Filled BCIDs are collected in the same *bunch train*, with many empty bunch spaces between trains, but none within them.

The beams only collide in dedicated collision points, where the beam paths intersect and around which detectors are built. The LHC has 8 such possible points, out of which 4 host large[3] experiments: ATLAS [2], ALICE [3], CMS [4] and LHCb [5]. While ALICE and LHCb specialise in heavy ion- and *b*-physics respectively, ATLAS and CMS were both built to be general multi-purpose experiments, designed to both discover the *H* boson and, if possible, BSM phenomena.

---

[1] As this work focuses on the proton collision data, the lead ion beams will not be discussed further.

[2] Apart from an initial, comparatively small 50 ns dataset.

[3] In addition there are smaller experiments, which will not be detailed here.

© Springer International Publishing AG 2017
L.K. Bryngemark, *Search for New Phenomena in Dijet Angular Distributions at √s = 8 and 13 TeV*, Springer Theses, DOI 10.1007/978-3-319-67346-2_3

**Table 3.1** Some design goal specifications for the LHC [1], along with actual performance by the end of 2012 and 2015

|                                          | Design        | 2012                | 2015                |
|------------------------------------------|---------------|---------------------|---------------------|
| Beam energy (TeV)                        | 7             | 4                   | 6.5                 |
| Dipole magnetic field (T)                | 8.33          | ~6.3                | ~8.0                |
| Dipole cooling medium                    | Liquid He     |                     |                     |
| Dipole temperature                       | 1.9 K         |                     |                     |
| Peak luminosity in ATLAS (cm²/s)         | $10^{34}$     | $7.3\times10^{33}$  | $5.0\times10^{33}$  |
| Number of protons per bunch              | $1.15\times10^{11}$ | $1.7\times10^{11}$ | $1.2\times10^{11}$ |
| Number of proton bunches                 | 2808          | 1374                | 2244                |
| Number of bunch places                   | 3564          |                     |                     |
| Bunch spacing (ns)                       | 25            | 50                  | 25                  |
| Stored beam energy (MJ)                  | 362           | 150                 | 280                 |
| Expected luminosity lifetime             | 14.9 h        | –                   | –                   |
| Minimum turnaround time                  | 70 min        |                     |                     |
| Expected average turnaround time         | 7 h           | –                   | –                   |
| Integrated luminosity/year (fb$^{-1}$)   | 80–120        | 20.3                | 3.6                 |

## 3.1   Collider Kinematics

The protons in the beam each carry the beam energy, presently 6.5 TeV.[4] As two protons collide, the energy in the centre-of-mass frame, the centre-of-mass energy, is $\sqrt{s} = 2 \times 6.5 = 13$ TeV. Since the two protons have the same mass and same kinetic energy, they have opposite but equal magnitude momenta $|\vec{p}|$,[5] and the proton-proton ($pp$) centre-of-mass frame and the detector or laboratory frame coincide. But, as we have learned, protons are composite particles, and in an LHC collision the energy is high enough to resolve the quark and gluon constituents. These constituents each carry a fraction of the proton momentum, which we denoted with (Bjorken) $x$. Thus, colliding partons may not have equal and opposite momenta, which means that the colliding centre-of-mass frame may differ from the detector frame. Here the concept of rapidity $y$ comes in handy:

$$y = \frac{1}{2} \ln \left( \frac{E + p_z}{E - p_z} \right) \tag{3.1}$$

using the energy $E$ and longitudinal momentum $p_z$ of a particle. "Longitudinal" and "transverse" directions are taken with respect to the beam axis at the collision point. A collision system with unbalanced incoming longitudinal momenta in the

---

[4]The proton mass energy is negligible: ~1 GeV, 10000 times smaller than the kinetic energy.

[5]"Coincide" is a simplification: in reality a small crossing angle gives the $pp$ frame a non-zero transverse component in the lab frame.

lab frame will acquire a rapidity *boost* and as a whole move longitudinally along the beam direction. One can show that rapidity is additive, implying that the rapidity $y$ as measured in the detector frame is related to that of the collision centre-of-mass frame $y_{CM}$ through the simple transformation

$$y = y_{CM} + y_B \tag{3.2}$$

where we have introduced the boosted system rapidity with respect to the detector frame, $y_B$.

Given that the initial state momentum is not clearly known, the quantity of interest at a hadron collider tends to be the transverse momentum, $p_T$. Since the protons move along the beam axis and the partons inside have relatively small intrinsic transverse motion, the initial state $p_T$ is considered to be 0. Being a vectorial quantity, linear momentum is conserved also component wise, so we immediately know two things: that all transverse momentum must have been transferred in the collision, and that the total transverse momentum of all outgoing particles should sum up to 0. This makes $p_T$ extremely useful for the hadron collider physicist.

We can now express the four-momentum of a massless particle as

$$
\begin{aligned}
p^\mu &= (E, p_x, p_y, p_z) \\
&= (p_T \cosh(y), p_T \sin(\varphi), p_T \cos(\varphi), p_T \sinh(y)),
\end{aligned}
\tag{3.3}
$$

where all quantities are experimentally observable.

### 3.1.1 Luminosity and Probability

In nuclear and particle physics, probability is often referred to in terms of cross section ($\sigma$).[6] To discover the rare processes we haven't seen yet but we do hope are technically possible, we need a large number of collisions. The key concept here is instantaneous luminosity, $L$, defined through the rate $R$ of events of some type and the cross section for them to happen:

$$R = L \cdot \sigma \tag{3.4}$$

which in a time interval $\Delta t$ gives a number of events

$$N = \sigma \int_t^{t+\Delta t} L dt \tag{3.5}$$

which introduces the concept of integrated luminosity.

---

[6]*cross section*, in units of *barns*, b, or cm$^{-2}$: in some sense an area, a geometrical image of how likely it is to hit something. 1b $= 10^{-28}$m$^2$.

For a given cross section for a new phenomenon, a higher integrated luminosity will thus increase the chances for discovery. Conversely, if the integrated luminosity is known, information on the cross section can be deduced from the number of events observed. $N$ and $L$ are experimental quantities, while $\sigma$ is a theory parameter, containing the information about the modelling of the physics process—the information we are really interested in. Knowing the integrated luminosity is thus the key to make theory interpretations of an experimental event count, and it is given in units of $fb^{-1}$ or $pb^{-1}$ on almost every data figure made public by a collider experiment.

## 3.2  Collider Data Taking

This section introduces some data taking nomenclature, along with the specifics of selection of data to record.

In order to understand the data taking at a collider, it is useful to keep a couple of quantum mechanical facts in mind:

1. anything that can technically happen, will happen eventually (it's a matter of probability)
2. we already know about the things that happen quite often (it's a matter of probability)

As we shall see, these principles govern the design of accelerators as well as the data acquisition strategy of the experiments.

## 3.3  The LHC/Beam Conditions

Apart from an extended accelerator shutdown period in the winter, the experiments are continuously taking data. In practice this means that proton bunches are injected into the LHC (a "fill" starts), accelerated up to collision energy, and then the beams are focused in the accelerator optics to optimise the bunch geometries with respect to beam losses and collisions in the collision points. Once the beam properties are optimised, "stable beams" are declared, and collision data taking can begin. Data are collected over a period of time where beam and detector conditions are stable. This is called a run, and maximally lasts for the duration of a fill (but often shorter). As seen in Table 3.1, one fill can often last more than 15 h.

## *Triggering*

In the LHC beam conditions as of 2015, approximately every 25 ns there is a bunch crossing, and in each bunch crossing, there are multiple proton collisions.[7] If the signals from every collision were to be read out, that corresponds to a rate of 40 MHz. In reality, ATLAS only writes collision data to tape at a rate of 1 kHz. This reduction[8] by a factor $10^5$ is not random, but based on selecting collisions according to certain criteria, viz. a signal from a certain amount of energy deposited in a certain sub-detector, and combinations thereof. This is called a *trigger*. At every bunch crossing, the detector signals are written to a buffer, and read out only if a trigger has "fired". A trigger is based on signatures such as two energy depositions compatible with electrons passing through the detector, above a certain energy threshold, or, an energy deposition compatible with a highly energetic hadronic jet. In principle, the outcome of one given collision can fire multiple triggers. The triggered and recorded signal from a bunch crossing is called an *event*. In the subsequent data analysis, the trigger decision is used to select the events that are interesting with respect to the phenomenon one is interested in.

As mentioned previously, the already known phenomena are the ones occurring more frequently. For instance, the energy distribution of jets is steeply falling, meaning, that high-energy jet events are very rare compared to low-energy ones. This implies that in a range of single jet triggers at different energy thresholds, the low-threshold ones fire much more often than the ones at higher threshold. But, the high energy events are often of much higher interest. In order to reduce the relative triggering rate of less interesting events, allowing more bandwidth for the more ones considered more interesting, it is very common to apply a *prescale factor*. A prescale factor $N_p$ means that only one in every $N_p$ triggered events is actually recorded. A trigger with prescale factor $N_p = 1$ is called un-prescaled. The real rate of the triggering process is recovered by multiplying with the prescale factor, which in effect is a weight. The statistical precision is however smaller than that achieved by recording the full set of events. This means a loss of sensitivity to new phenomena at lower $p_T$, implying that it will take more data (longer time) to discover them. Using un-prescaled triggers only gives full sensitivity, but limited to the higher $p_T$ regime.[9]

---

[7]On average 20.7 in 2012 (50 ns bunch spacing), and 13.5 in 2015.

[8]Going from 40 MHz to 1 kHz means dismissing 99.9975% of the data.

[9]An elegant solution to this experimental trade-off is to read out a minimal amount of information from each event, which allows storing these at a higher rate. The challenge is ensuring that the reconstruction of these jets does not suffer from the loss of information from for instance the tracker. Far from being my idea, I still venture to say that with higher luminosities ahead of us, this type of "trigger-level" analysis and fast reconstruction of objects at trigger level is the way forward to retain sensitivity to phenomena in the sub-TeV scale, without requiring associated production of objects whose dedicated triggers have a lower prescale.

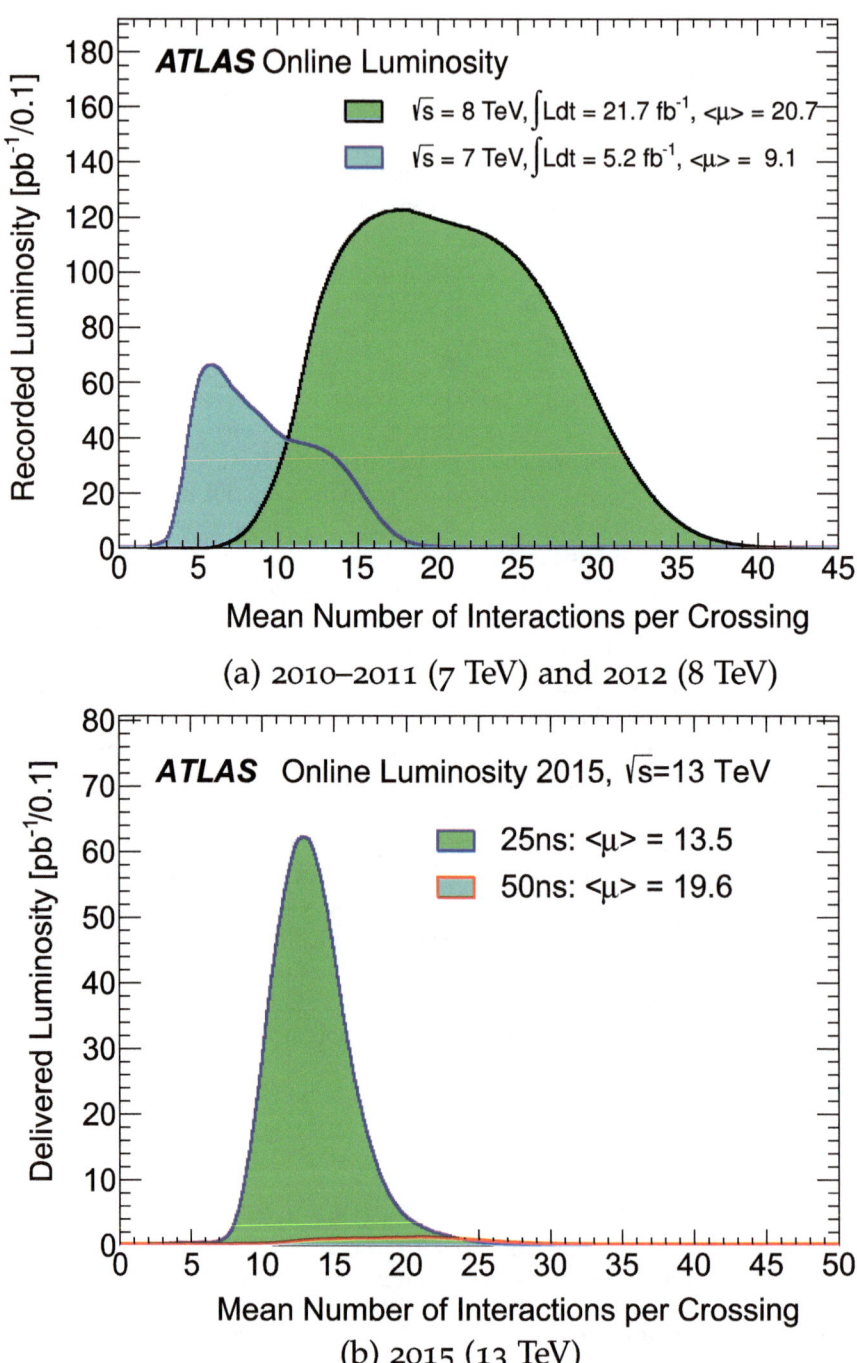

**Fig. 3.1** The distribution of the average number of simultaneous interactions per bunch crossing for data taking in **a** 2010–2012 and **b** 2015

## 3.4  Pile-Up

In Sect. 3.1.1 we introduced the concept of luminosity and linked that to the probability to observe a new, probably rare, phenomenon. In order to maximise the instantaneous luminosity, and thus quick discovery potential, the LHC operates in a mode of several simultaneous proton-proton interactions in a given bunch crossing. Most of these are processes that are already well known, but occasionally, a process that triggers the readout happens. The full event is then recorded, along with the overlaid activity from the simultaneous interactions. This is referred to as pile-up, which is a common term for when multiple signals get overlaid in detector readout, and generally this relates to the relationship between the rate of interactions and the signal collection and readout time. I stress that at the current typical LHC conditions, the overlaid events are *simultaneous* with the triggering collision within a time interval very much smaller than the bunch spacing, and events without pile-up are extremely rare, as is seen in Fig. 3.1. Since there is no way of avoiding collecting all this overlaid signal at once, techniques to recognise the triggering collision and correct for the signal from the rest have to be devised.

Figure 3.1b also shows the difference in the average number of simultaneous interactions per bunch crossing, denoted by $\langle \mu \rangle$, achieved with 25 and 50 ns bunch spacing, respectively. With fewer bunches, a larger number of *pp* collisions have to occur in the same bunch crossing to achieve the same instantaneous luminosity.

## References

1. L. Evans, P. Bryant (eds.), LHC machine. JINST **3**. ed. by L. Evans, S08001 (2008)
2. G. Aad et al., The ATLAS experiment at the CERN large hadron collider. JINST **3**, S08003 (2008)
3. K. Aamodt et al., The ALICE experiment at the CERN LHC. J. Instrum. **3**(08), S08002 (2008)
4. S. Chatrchyan et al., The CMS experiment at the CERN LHC. JINST **3**, S08004 (2008)
5. A.A. Alves Jr et al., The LHCb detector at the LHC. JINST **3**(08), S08005 (2008)

# Chapter 4
# The ATLAS Experiment

The ATLAS detector, drawn to scale in Fig. 4.1, is an impressive beast in many ways. To the visitor approaching in the pit 100 m below ground, already the size of this gradually emerging lying cylinder, 25 m high and 45 m long, inspires awe. It comprises several sub-detectors, each employing their own technique to contribute their piece to the puzzle of particle identification, and trajectory and energy measurement. These have been designed and built over decades, in collaboration between different institutes across different continents, and are operated day and night in the same spirit. Reading out, storing, and analysing the large amount of data it produces requires an equally large effort in design and collaborative operations. In this chapter, I will summarise the steps needed from proton collision to the data analysis that the rest of this book is devoted to. It has to be a selective description, and for more complete descriptions of ATLAS, the reader is referred to Ref. [1].[1]

## 4.1 Coordinate System

ATLAS uses a right-handed coordinate system with its origin in the centre of the detector at the nominal interaction point (IP) and the $z$-axis directed along the beam pipe.[2] The positive $x$-axis points from the IP to the centre of the LHC ring, and the $y$-axis points upward. Cylindrical coordinates $(r, \varphi)$ are used in the transverse plane, $\varphi$ being the azimuthal angle around the $z$-axis. The Pseudo rapidity is defined in terms of the polar angle $\vartheta$ as $\eta = -\ln\tan(\vartheta/2)$ with $\vartheta$ measured with respect to the $z$-axis. Pseudo rapidity is a convenient concept when discussing coverage with respect to the beam axis.

---

[1]For coming upgrades see for instance Refs. [2, 3].

[2]This differs from the beam axis: the beams collide at a slight angle to avoid interactions upstream.

© Springer International Publishing AG 2017

L.K. Bryngemark, *Search for New Phenomena in Dijet Angular Distributions at √s = 8 and 13 TeV*, Springer Theses, DOI 10.1007/978-3-319-67346-2_4

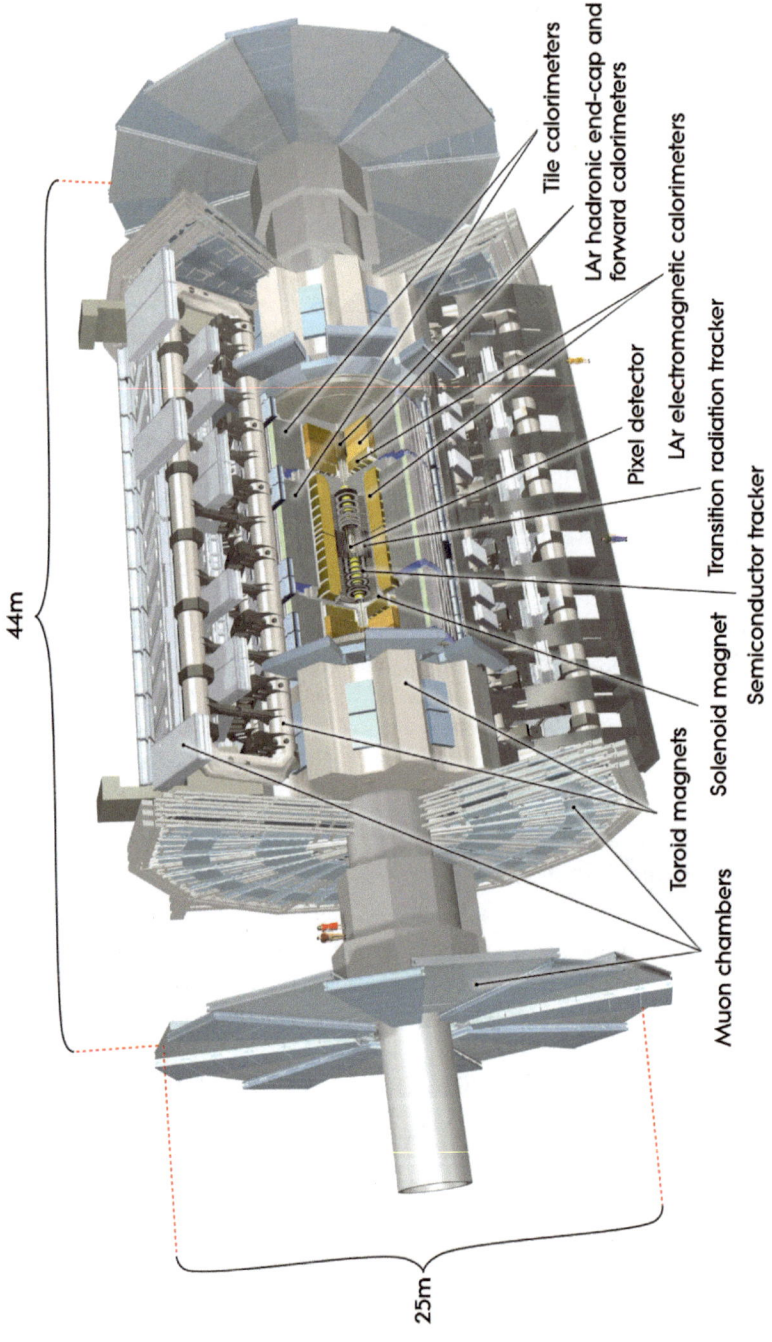

**Fig. 4.1** The ATLAS detector, with subdetectors indicated [4]

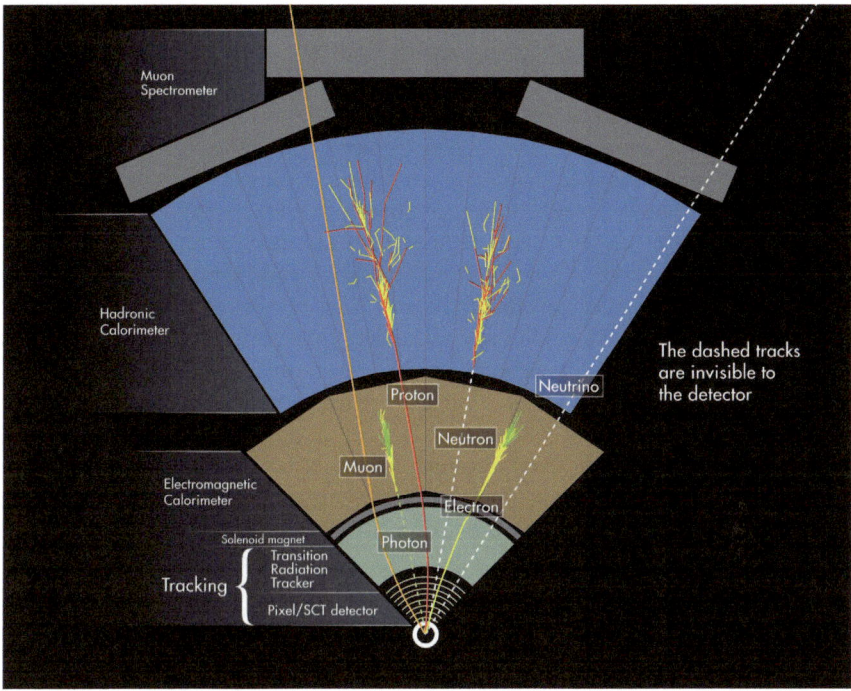

**Fig. 4.2**   The principle of particle identification through interactions with a cross-sectional wedge view of the ATLAS detector [4]

## 4.2   Collider Particle Detectors: The Onion Design

The philosophy of an experiment designed for the discovery of something generally new and unknown is to measure everything—to not let anything escape. Now this is of course not physically possible with the knowledge we have—for instance, as mentioned in Sec. 2.2.1, neutrinos easily traverse entire planets without interacting at all. The next best thing, is then to make sure that the rest of the known particles will interact in the detector, such that upon creation, their properties can be carefully measured, and the remaining details deduced. To achieve this, the subdetectors are ordered radially from the interaction point—where particles are created and travel outwards—in such a way that finer details can be resolved in the beginning, and in the outer layers, the particle energies are measured by complete absorption—in effect, destruction of all remaining information!

The basic principles are illustrated in Fig. 4.2. Charged particles, like electrons and protons, interact with the tracker material, while neutral particles like photons and neutrons don't. Photons do shower in the electromagnetic calorimeter, while neutrons only interact hadronically. Combining the signal from the different subdetectors, one can thus distinguish these particle types. Muons barely interact with any of the material, and are thus the only particle to make it all the way out to the muon spectrometer. Neutrinos escape undetected as they don't interact with the material at all. Their presence can be deduced from non-conservation of total transverse momentum in a collision (so-called missing transverse energy, $E_T^{miss}$).

It is an overall feature that there is finer segmentation of the detector closer to the centre. Apart from the pure geometrical fact that the density of produced particles will decrease radially as $1/R^2$, this is the result of a trade-off between precision measurement and the cost of finely segmented read-out. Electronics require room, feed-throughs, cracks in the detector coverage. They introduce material and heat load. Cooling introduces further material and cracks. Un-instrumented ("dead") material in the detector reduces the resolution in the energy measurement. Cracks introduce regions where particles are poorly measured or even escape the detector. Fine segmentation is thus only used where it is most needed.

The rest of this chapter will describe ATLAS's take on this design philosophy. It will be a maze of acronyms, but be patient; hopefully it's useful for later reference. Throughout this section, keep in mind that even when only one detector is mentioned, the symmetry in $\eta$ entails that there is one of each detector type in the *end-caps* located on each side of the central *barrel* region.

## 4.3  The ATLAS Detector Subsystems

While most of the information in this chapter is available in Ref. [1], ATLAS continuously evolves, with many of its institutes involved in detector research and development in parallel to data taking. During the long shutdown between Run1 and 2, a new vertexing detector, the Insertable B-Layer (IBL) [5, 6] was installed closest to the beam pipe.

### 4.3.1  Magnets

ATLAS uses two magnet systems: an outer air-core toroid system, and a thin 2 T superconducting solenoid surrounding the inner detector containing tracking detectors. Figure 4.3 shows a sketch of the magnet geometry.

A magnetic field is crucial for momentum determination of charged particles, as the bending radius of a charged particle trajectory is proportional to the momentum perpendicular to the magnetic field direction. Since the solenoid and beam axes are aligned, this corresponds to determining the $p_T$ of the particle. However, at high $|\eta|$, only a fraction of a particle's momentum is perpendicular to the field from the solenoid. The toroid system is intended for determining muon momentum with high precision and is arranged such that the field is mostly perpendicular to the muon trajectories even at large $|\eta|$, thus compensating the limitation of the solenoid. The open structure reduces the material in front of the muon spectrometers to a minimum, minimising the effect of multiple scattering that would deteriorate the resolution.

### 4.3.2   The Inner Tracker: Silicon Strips and Pixel Detector

Given that particles are produced in the interaction point, the track density is very high in the innermost detectors, requiring excellent spatial resolution of signal for momentum and vertex reconstruction. This is achieved with highly granular silicon tracking detectors, with concentric cylindrical geometry in the barrel region and perpendicular disks in the end-cap. The overall layout is seen in Fig. 4.4. Tracking extends to $|\eta| = 2.5$ in total, with the highest granularity detectors, the pixel and IBL detectors, located around the vertex region close to the IP. A typical track crosses eight strip layers (making four space points) and three pixel layers, giving position coordinates in 3D: the radius $R$, $\varphi$ and $z$. The intrinsic accuracy in $R - \varphi$ is $10\,\mu$m in pixel, and $17\,\mu$m in the strips.

The most important job for the innermost detectors is vertex reconstruction. In an environment with high particle multiplicity and several proton collisions in the same bunch crossing, high precision reconstruction of the primary vertex, indicating the primary interaction point, is absolutely mandatory. Furthermore, for the identification of heavy quarks such as $b$ or $c$-quarks, whose relatively long lifetime allows

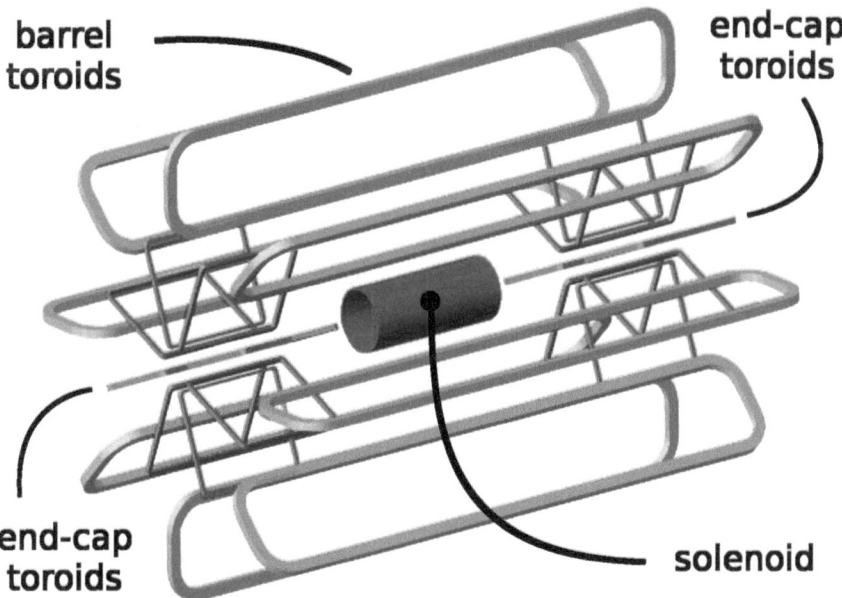

**Fig. 4.3**  The ATLAS magnet system [7]

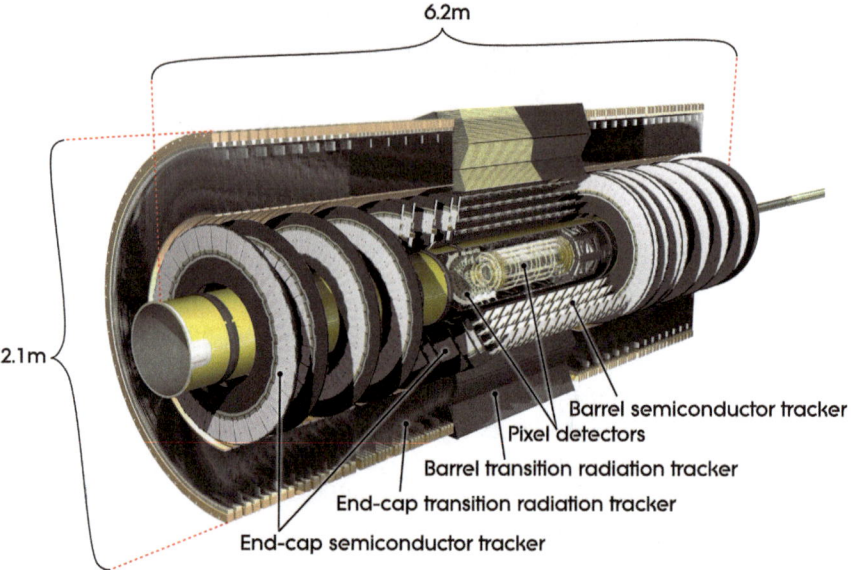

6.2m

2.1m

Barrel semiconductor tracker
Pixel detectors
Barrel transition radiation tracker
End-cap transition radiation tracker
End-cap semiconductor tracker

**Fig. 4.4** A zoomed-in cut-away view of the inner detector [4]

travelling a detectable distance before decaying, reconstruction of displaced secondary vertices, and determining their degree of pointing back to the primary vertex, is equally necessary.

As mentioned, the inner detector has recently been complemented with an additional pixel layer, the IBL. It improves the vertexing capabilities of ATLAS and supplements the previously innermost layer which has been, and will increasingly be, exposed to much radiation. Loss of coverage in these detectors would dramatically deteriorate vertex reconstruction as well as $b$-tagging.

### 4.3.3  The Transition Radiation Tracker

The TRT is a gaseous ionisation detector, with drift chambers made from thin cathode tubes (or *straws*) containing mostly Xe or Ar, and $CO_2$ for avalanche quenching. It surrounds the inner silicon tracker detectors and has a similar structure, as can be seen in Fig. 4.4: the straws are parallel to the $z$-axis in the barrel and radial to it in the end-cap. Signal is read out from thin anode wires in the straw centre. It is specialised in distinguishing electrons from other particles, using the *transition radiation* emitted when they cross a boundary between two media of different refractive indices. All charged particles emit such radiation, but the energy radiated is proportional to the

Lorentz factor $\gamma$ which for a given energy[3] is higher for lighter particles. Since electrons are much lighter than all other charged particles, this radiation turns out to be a good discriminant—especially against charged pions which could otherwise introduce ambiguities in case of insufficient information from other subdetectors. The transition radiation is provoked by radiators placed between the straws, and the X-rays thus produced are absorbed in the Xe, making a large ionisation signal.

Apart from its electron identification capacity, the TRT provides a large number of 2D points in $R - \varphi$ along a track, with an intrinsic accuracy of $130\,\mu$m per straw. When combined with the 3D information from the inner tracker, they extend the track lever arm, enhancing the spatial as well as momentum resolution.

### 4.3.4 Calorimetry

The topic of calorimetry is of particular importance to much of the work described in this thesis. There will be a complete chapter dedicated to the calorimeters (Chap. 5) in the next part of the thesis. Here only the main features of the ATLAS calorimeters are mentioned.

The task of a calorimeter is to fully contain and measure the energy of an incident particle. Thus, it also absorbs the particle and no further measurement on it is possible. Good absorption prevents leakage out into the muon detectors.

An overview of the ATLAS calorimeter system is shown in Fig. 4.5. The combined ATLAS calorimeter systems provide near-hermetic electromagnetic and hadronic coverage out to $|\eta| = 4.9$. All ATLAS calorimeters have longitudinal and lateral segmentation to provide directional information. The central parts of the ATLAS calorimeters have high granularity to enable pointing back to the primary vertex. The calorimeter is also extensively used for triggering.

### 4.3.5 Muon Spectrometers

The outermost detector system of the big ATLAS cylinder is the muon system. It consists of chambers placed in three layers, cylindrical in the barrel region and in perpendicular planes in the end-cap region. They employ a range of charged particle detection techniques in subdetectors: drift tubes and cathode strip chambers for precision tracking, and fast resistive plate and thin gap chambers for triggering

---

[3]With $\beta = v/c$, the particle's velocity $v$ expressed as a fraction of the speed of light $c$, we define $\gamma$ as

$$\gamma = \frac{1}{\sqrt{1 - \beta^2}} \tag{4.1}$$

and note that at fixed energy, lighter particles travel faster. It is clear that $\gamma$ increases very steeply at high $\beta$, making the impact of even small changes in mass quite visible.

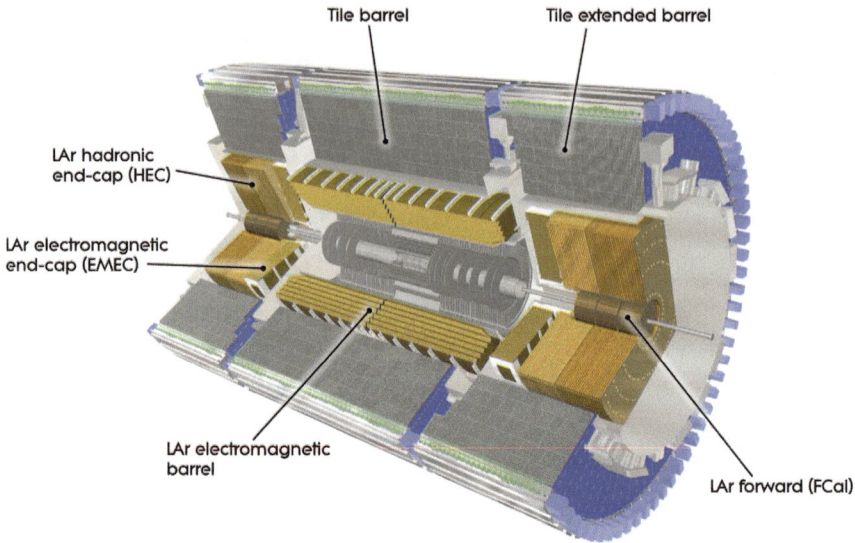

Tile barrel

Tile extended barrel

LAr hadronic
end-cap (HEC)

LAr electromagnetic
end-cap (EMEC)

LAr electromagnetic
barrel

LAr forward (FCal)

**Fig. 4.5**  A cut-away view of the calorimeters [4]

and second trajectory coordinate. The only hint that signal in these chambers comes from a muon is the simple fact that other charged particles are rarely expected to make it as far out in the ATLAS detector system.

Given the large scale of these detectors, relative alignment becomes crucial for an accurate momentum reconstruction. For instance, the drift tube alignment is continuously monitored using about 12,000 precision-mounted alignment sensors, which optically detect deviations from straight lines. As for all tracking detectors, the final alignment is done from the reconstruction of tracks measured in the detectors themselves.

### 4.3.6  LUCID

Closest to the beam pipe, 17 m from the interaction point, the luminosity measurement detector LUCID (LUminosity measurement using Cerenkov Integrating Detector) sits. It detects the Cerenkov radiation produced in quartz windows as charged particles from the collision debris pass through it. It is in principle a particle counting device, using the approximate proportionality between particle multiplicity and the number of interactions to measure the instantaneous luminosity of the proton beam collisions. For an excellent account of the methods to do this, see Ref. [8]. Here, all we need to know is that the luminosity is deduced from the number of particles measured and the cross section for inelastic proton collisions. It is averaged over a *lumiblock*—

some period of time extending across several bunch crossings, typically of order one minute.

### 4.3.7 More Forward: ALFA and ZDC

ALFA (Absolute Luminosity for ATLAS) and the ZDC (Zero-Degree Calorimeter) are located at 240 and 140 m from the interaction point, respectively. ALFA uses Roman pots to measure the proton-proton cross section, by counting forward protons[4] as close to 1 mm from the beam. The ZDC measures the neutral component of beam remnants (spectators), which travel in a straight line while the beam bends, to measure centrality in lead ion collisions. It is also used for detection of diffractive processes and minimum-bias triggering.

## 4.4 Detector Simulation

Apart from collision data, ATLAS relies heavily on simulated collisions to make predictions based on the SM. The MC generators described in Sect. 2.7 can create detailed *MC truth*[5] or *particle level* records of the particles produced in a simulated collision, but they don't take any detector effects into account. These effects can be everything from pure acceptance (or coverage) effects to energy loss in detector material or stochastic effects affecting the position resolution, and they are present all the time in the real data. To correct for this, simulated particles are "propagated" through a GEANT4 [9] detector simulation, where the passage of particles through matter is simulated in detail, based on the ATLAS geometry and material. This results in *reconstructed* or *detector level* distributions.

## 4.5 ATLAS Conditions

From collision to recording and analysis of data, there are still a few steps. Those common to all ATLAS data taking will be outlined here.

---

[4]The optical theorem relates the forward scattering amplitude of a process to its total cross section.

[5]The concept of "truth" is often used when assessing the impact from experimental conditions on an observable.

### 4.5.1   Trigger System

ATLAS uses trigger system split into a fast trigger system, Level-1 (L1), implemented
in hardware and making the initial selection of events, and a slower system using the
L1 decisions as input. In Run1 these were passed to the software triggers at Level-2
and the Event Filter (EF) level, which used Regions Of Interest to identify for instance
localised high-energy deposits in the calorimeter, using a jet finding algorithm. In
Run2 the software level was reduced to one High-Level Trigger (HLT). In Run1 the
EF maximum accept rate was 400 Hz while in Run2 the HLT is capable of a rate of
1 kHz.

### 4.5.2   Data Quality

As mentioned in Sect. 4.3.6, the average luminosity is measured in lumiblocks of
1–2 min. Generally, data quality stamps are also assigned on a lumiblock basis (for
instance, high-voltage stability in the LAr calorimeters). The data taking lumiblocks
where all detector conditions are understood and smooth are listed in a *Good Runs
List* (GRL) applied as part of the event selection in every data analysis.[6]

### 4.5.3   Data Processing

The outcome of both collision data taking and simulation of physics processes propagated through a detector simulation, is signals in the channels of the detector. The
detector signals from a triggered event are digitised *online* (in real time) and through
existing mapping from readout channel to geometry, reconstruction of for instance
tracks and calorimeter energies can begin. The RAW detector data are here trans-
formed to event level data, which amounts to a large reduction in size. In particular,
storing a handful of track parameters requires much less space than storing ionisation information and coordinates in 3D for each hit. This size reduction continues
as energy deposits in calorimeter cells are combined to form physics objects such as
jet or electron candidates. The final data format used for analysis, presently in the

---

[6]The requirements on the detector differ between analyses; so do the GRLs.

form of a *DAOD* (for Derived Analysis Object Data) often only includes the physics object candidates of interest to the analysis, along with trigger decision information and event-level information used for calibration and matching to the GRL, for instance. This amounts to both a reduction in size and a loss of detail available to the user, highlighting the need for several types of DAODs. Although substantially reduced in size, the final data set can still amount to several TB of data, which is too large for the individual user to store. Much of the analysis and the previous steps are thus done on common infrastructure and with distributed computing—simply put, the much more lightweight analysis code is sent to where the data are stored instead of the other way around. The event selection applied in the user code often reduces the data set size drastically, giving final files which are easily manipulated and stored on a regular laptop.

# References

1. G. Aad et al., The ATLAS experiment at the CERN Large hadron collider. J. Instrum. **3**, S08003 (2008)
2. ATLAS Collaboration, New small wheel technical design report. Technical Report CERN-LHCC-2013-006. ATLAS-TDR-020. Geneva: CERN (2013)
3. ATLAS Collaboration, Fast TracKer (FTK) technical design report. Technical Report CERN-LHCC-2013-007. ATLAS-TDR-021. Geneva: CERN (2013)
4. ATLAS Experiment © 2014 CERN (2016), http://www.atlas.ch/photos
5. ATLAS Collaboration, ATLAS insertable B-layer technical design report. Technical Report CERN-LHCC-2010-013. ATLAS-TDR-19. Geneva: CERN (2010)
6. ATLAS Collaboration, ATLAS insertable B-layer technical design report Addendum. Technical Report CERN-LHCC-2012-009. ATLASTDR- 19-ADD-1. Addendum to CERN-LHCC-2010-013, ATLASTDR- 019. Geneva: CERN (2012)
7. J.J. Goodson, Search for supersymmetry in states with large missing transverse momentum and three leptons including a Z-Boson. Presented 17 Apr 2012. Ph.D. thesis. Stony Brook University (2012)
8. A. Floderus, Luminosity determination and searches for supersymmetric sleptons and gauginos at the ATLAS experiment. Presented 30 Jan 2015. Ph.D. thesis. Lund: Lund University (2014)
9. S. Agostinelli et al., GEANT4: a simulation toolkit. Nucl. Instrum. Method A **506**, 250–303 (2003)

# Part II
# Jets

Having settled that partonic interactions, and notably hadronic final states, can be observed through the emergence of jets, we must now proceed to establishing a procedure for identifying a jet from our collision. This is a rich topic, as it encompasses the full range from our theoretical understanding of the mechanisms producing a jet, over the measurement of energy in our detectors and the difficulties associated with identifying the origin of that energy, to some clever assignment mapping some of it to our theoretical concept of a hadronic shower originating from an energetic parton. Luckily, it is also a crucial concept for the understanding of my research, and I will have the pleasure of expanding on it at length!

# Chapter 5
# Calorimetry

This section will briefly describe the principles of energy measurement in calorimeters, as well as the algorithms used for identifying signal corresponding to energy depositions from interacting particles.

## 5.1 Electromagnetic Calorimetry

The electromagnetic (EM) calorimeter detects photons and any charged particle through the energy depositions from the EM showers resulting when these particles traverse a dense medium. The thickness of a calorimeter is often expressed in *radiation lengths* $X_0$. It is the mean distance travelled before an electron loses $1/e$[1] of its energy through emitting *bremsstrahlung*. Conversely, it is $7/9$ of the mean free path before a photon above threshold undergoes pair production. These two processes lead to an EM cascade or *shower* of electrons and photons. The radiation length is a characteristic scale of the EM shower evolution, but corresponds to a material specific actual length. Similarly, the transverse dimension of the shower is given by the Molière radius $R_M$, which relates to the radiation length approximately as $R_M = \frac{21\,\mathrm{MeV}}{E_c}$, where the critical energy $E_C = \frac{500\,\mathrm{MeV}}{Z}$, depends on the atomic number $Z$ of the material. This radius gives the material specific shower position resolution: on average, 90% of the shower energy is contained within a cylinder with this radius centred on the shower.

The cross section for EM processes is proportional to $Z^2$. The material choice tends to reflect this, as a higher cross section will reduce the actual length corresponding to $1X_0$ as well as $R_M$. Another consideration is material transparency to or readout of the energy depositions. This leads to two main technologies:

- homogeneous scintillating crystals, with at least one high-$Z$ element, which can be read out using photomultipliers, or

---

[1] $e$ here is the base for the natural logarithm: $e \approx 2.71828$.

© Springer International Publishing AG 2017
L.K. Bryngemark, *Search for New Phenomena in Dijet Angular Distributions at $\sqrt{s}$ = 8 and 13 TeV*, Springer Theses, DOI 10.1007/978-3-319-67346-2_5

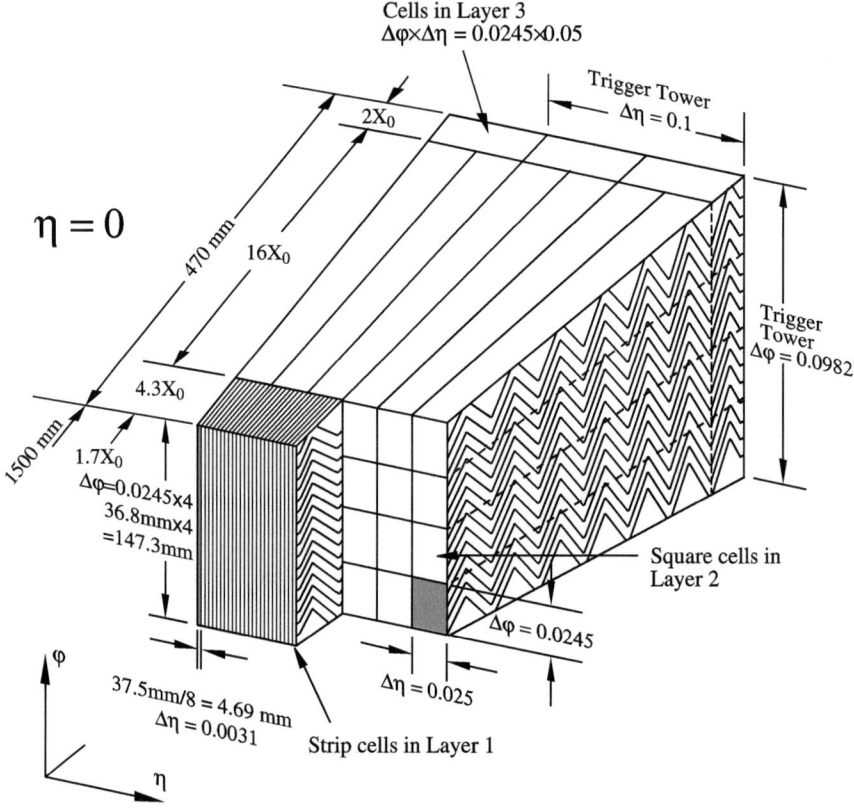

**Fig. 5.1**  A sketch of the LAr geometry in an EMB module, showing the lateral and longitudinal segmentation in layers of radially coarser granularity [1]

- sampling calorimeters where a high-$Z$ absorber provokes energy loss, while an interleaved active medium is responsible for its detection via ionisation or scintillation.

The choice of homogeneous/sampling calorimeter applies also to hadronic calorimetry, albeit with different optimal material choices; ATLAS uses only sampling calorimeters.

### 5.1.1   Liquid-Argon Electromagnetic Calorimeter

The Liquid-Argon (LAr) EM calorimeter is a layered detector, using lead[2] absorber plates and liquid argon as active medium producing ionisation. It is divided in Electro-

---

[2]Pb: $Z = 82$.

magnetic Barrel (EMB) and End-Cap (EMEC) parts, positioned outside the solenoid and housed in their own cryostats.[3] The absorber plates and electrodes are arranged in an accordion shape as illustrated in Fig. 5.1. The accordion shape provides full $\phi$ coverage and comparatively fast readout. To compensate for energy absorption in the solenoid, a thin pre-sampler layer of instrumented argon is located between the solenoid and the central parts of the EM calorimeter. The EM calorimeter is more than $22X_0$ thick, to ensure minimal leakage of electromagnetic energy.[4]

In $|\eta| < 2.5$, the region corresponding to the inner detector, the innermost layer has very fine granularity. Apart from allowing for precise measurement of photons and electrons, it enables distinction between neutral mesons (typically $\pi^0$) and photons. The latter is in fact a matter of separating showers from single photons, from those stemming from two collinear photons[5] from neutral meson decay. The small shower separation requires high granularity. Above all, from the first and second layers, this region has pointing capabilities to identify the vertex associated with photons, which don't produce tracks in the inner detector.

## 5.2 Hadronic Calorimetry

Hadronic energy loss is based on nuclear rather than atomic interactions. These include e.g. spallation, neutron capture and nuclear recoil. The secondary hadrons produced give rise to hadronic cascades as long as they carry enough energy. Hadronic cascades generally have relatively few high-energy particles compared to EM showers, and thus have large fluctuations. Eventually the low energy transfer reactions will remain undetected. For an electrically neutral hadron like a neutron, hadronic interactions is the only way to detect it. While the charged hadrons interact electromagnetically, they too may well produce neutral particles through nuclear interactions upstream. While most low-energy nuclear interactions such as recoils are lost, a fair fraction of the high-energy interactions lead to $\pi^0$ production which subsequently leads to EM showers in the hadronic calorimeter as the $\pi^0$ decays to a photon pair. Similarly, nuclear de-excitation $\gamma$ rays lead to EM interactions.

The characteristic longitudinal hadronic energy loss length is the *absorption length* $\lambda_{int}$. On average, an incident hadron will have interacted after this length, and from its definition one finds that this is the mean distance required to reduce by a factor $1/e$ the number of incident relativistic hadrons travelling through a medium. This is typically $\mathcal{O}(10)$ times longer than the radiation length in a medium,[6] meaning that hadronic calorimeters generally need to be very thick. Thus, hadronic calorime-

---

[3] Argon boils at 87 K.

[4] Electromagnetic energy here refers to energy that could have been transferred in electromagnetic interactions.

[5] The opening angle decreases with the meson $p_T$. At $p_T \sim 80$ GeV the $\pi^0$ rejection still benefits from segmentation.

[6] Remember the 5 orders of magnitude between atomic and nuclear size mentioned in Chap. 1!

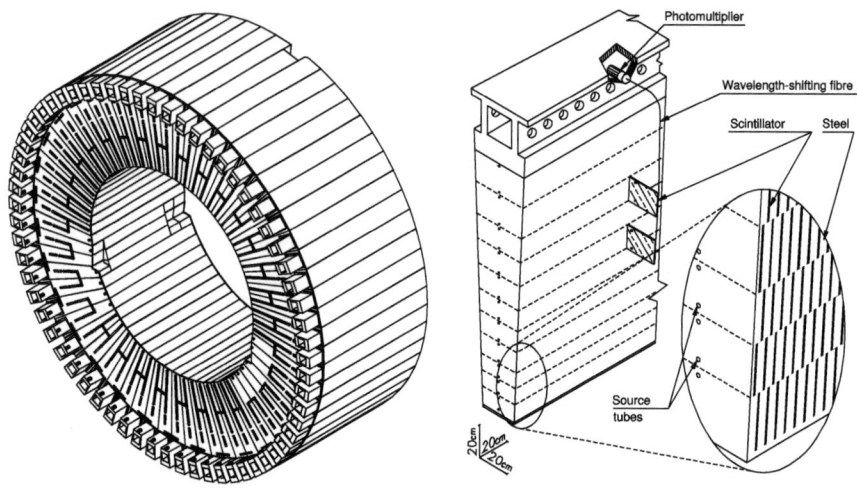

**Fig. 5.2** A sketch of a the Tile calorimeter, showing the overall layout and the scintillator tiles and steel plates as well as the readout fibres and photomultipliers. Particles enter from the centre (left) and bottom (right) [1]

ters form the outermost layer. The optimal choice of material is not necessarily the high $Z$ materials of EM calorimeters. For instance, very light atoms such as hydrogen are often advantageous for neutron detection.

The relative detection efficiency of hadronic and electromagnetic energy depositions in a hadronic calorimeter is denoted by $h/e$. From the discussion above, it is clear that the detection efficiency of hadronic interactions is often comparatively small. If $h/e = 1$, the detector is said to be compensating. This can be achieved by reducing the EM efficiency, increasing the hadronic efficiency (through the use of materials containing hydrogen) or by introducing a fissile material producing additional radiation after fast neutron capture.

The ATLAS hadronic calorimetry is non-compensating and makes use of two different technologies, covering different range in $\eta$.

### 5.2.1  Tile

The Tile calorimeter uses steel[7] absorbers and scintillating polystyrene tiles as active material, read out with photomultiplier tubes via wavelength shifting fibres bringing the light from UV to the visible range. Figure 5.2 illustrates this. It has a barrel and an extended barrel part, reaching out to $|\eta| < 1.7$. The average radial depth is $7.4\lambda_0$, and its main purpose is to measure the energy and directions of hadronic jets.

---

[7]Fe: $Z = 26$. Plastics are organic molecules, containing much hydrogen.

**Table 5.1** Energy resolution in the ATLAS calorimeters

| Calorimeter segment | Energy resolution $\sigma_E/E$ (%) |
|---|---|
| EMB | $\frac{10\%}{\sqrt{E}} \oplus 0.7$ |
| EMEC | $\frac{10\%}{\sqrt{E}} \oplus 0.7$ |
| HEC | $\frac{50\%}{\sqrt{E}} \oplus 3$ |
| FCal | $\frac{100\%}{\sqrt{E}} \oplus 10$ |
| Tile | $\frac{50\%}{\sqrt{E}} \oplus 3$ |

## 5.2.2   LAr Forward Calorimeters

The Hadronic End-Cap (HEC) and Forward Calorimeter (FCal) are LAr calorimeters, housed in the same cryostats as the EMEC. The HEC uses copper[8] absorber plates instead of lead, while the FCal uses both copper and tungsten plates: copper for EM calorimetry in the innermost layer and tungsten in the outer two for hadronic. The HEC has a depth of $\sim 12\lambda_0$ while FCal, sitting in a region of very large activity at small angles to the beam, has a total depth of $\sim 10\lambda_0$ and $\sim 200X_0$. The challenging radiation environment led to a coaxial electrode design, enabling smaller LAr gaps than in the barrel,[9] to avoid problems with ion build-up while retaining high material density. In addition, the signal is faster than in the barrel due to the shorter drift time in the LAr gap.

## 5.3   Resolution: Energy and Granularity

Not only is the task of the calorimeter to fully contain the energy from high-energy particles: as the name suggests, they also measure it. Unlike for tracking detectors, which rely on trajectory bending for momentum resolution, the calorimeter energy resolution gets better with incident energy. The relative energy resolution can be expressed as

$$\frac{\sigma}{E} = \frac{a}{E} \oplus \frac{b}{\sqrt{E}} \oplus c, \tag{5.1}$$

where the terms are in turn referred to as the *noise term*, the *sampling term*, and the *constant term*, respectively. We will discuss noise soon enough. The sampling term depends on material choice and thickness. The constant term is what dominates at high energy, and it is governed by the detector layout and geometry: how many radiation or absorption lengths and how uniform it is, if there are cracks and dead

---

[8]Cu: $Z = 29$.
[9]0.2–0.5 mm instead of 2 mm.

**Table 5.2** The $|\eta|$ coverage, typical granularity in $\eta$ and $\phi$ and number of readout cells of the ATLAS calorimeters, per segment and layer [1]

| | EM calorimeter | |
|---|---|---|
| | Barrel (EMB) | End-cap (EMEC) |
| $|\eta|$ coverage | 0–1.4 | 1.4–3.2 |
| Depth samples | | |
|   Presampler | 1 | – |
|   Calorimeter | 3 | 3 |
| Granularity $\eta \times \phi$ | | |
|   Presampler | $0.025 \times 0.1$ ($|\eta| < 0.8$) | – |
| | $0.003 \times 0.1$ ($|\eta| > 0.8$) | – |
|   Calorimeter | $0.003 \times 0.100$ | $0.003 \times 0.100$ ($|\eta| < 2.4$) |
| | $0.025 \times 0.025$ | $0.025 \times 0.025$ ($|\eta| < 2.4$) |
| | $0.025 \times 0.050$ | $0.025 \times 0.050$ ($|\eta| < 2.4$) |
| | | $0.050 \times 0.050$ ($|\eta| > 2.4$) |
| Readout channels | | |
|   Presampler | 7800 | 1500 |
|   Calorimeter | 100000 | 62000 |
| | LAr hadronic end-cap (HEC) | |
| $|\eta|$ coverage | 1.5–3.2 | |
| Depth samples | 4 | |
| Granularity $\eta \times \phi$ | $0.1 \times 0.1$ ($|\eta| < 2.4$) | |
| | $0.2 \times 0.2$ ($|\eta| > 2.4$) | |
| Readout channels | 5600 | |
| | LAr forward calorimeter (FCal) | |
| $|\eta|$ coverage | 3.1–4.9 | |
| Number of layers | 3 | |
| Granularity $\eta \times \phi$ | $\sim 0.15 \times 0.15$ | |
| Readout channels | 3500 | |
| | Tile hadronic calorimeter | |
| | Barrel | Extended barrel |
| $|\eta|$ coverage | 0–1.0 | 1.0–1.6 |
| Depth samples | 3 | 3 |
| Granularity $\eta \times \phi$ | $0.1 \times 0.1$ | $0.1 \times 0.1$ |
| | $0.2 \times 0.1$ (last sample) | $0.2 \times 0.1$ (last sample) |
| Readout channels | 5800 | 4100 |

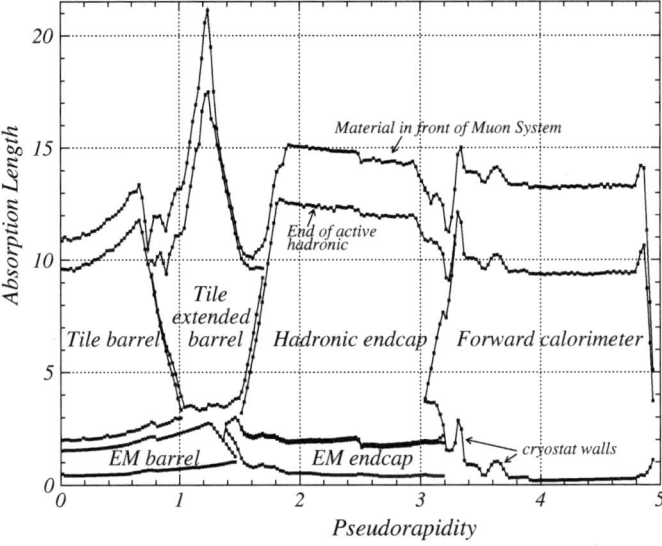

**Fig. 5.3**  The $|\eta|$ coverage and number of absorption lengths of the ATLAS calorimeter system [1]

(uninstrumented) material, etc. This term is reduced by interchannel calibration. In the other end of the energy range, the noise term dominates, but its importance decreases the fastest with $E$.

The incident particle energy resolution of the ATLAS calorimeter, neglecting the noise term which (as we will see later) is subject to large variations depending on the LHC pile-up conditions, is shown in Table 5.1.

The coverage and granularity of the calorimeters is detailed in Table 5.2. It shows that the cells grow larger farther from the IP. In Fig. 5.3, the coverage in both $|\eta|$ and material is represented visually in terms of number of absorption lengths of each calorimeter system.

## 5.4  Energy Measurements

The strength of the signal from a cell is a measure of the energy deposited in it. The conversion from ADC counts to signal current in $\mu$A, or charge in pC, is known from calibration, where a known charge is injected. The correspondence between signal current and energy is known from electron beam tests. After shaping, the amplitude of the pulse, which carries the energy information, is found through pulse fitting filtering algorithms. This will also give the timing of the pulse, used both to assign the signal to a BCID and for data quality purposes. The amplitude of the un-amplified

**Fig. 5.4** The amplitude of
the LAr raw triangular pulse,
drift current versus time,
overlaid with the that from
the bipolar shaping. 25 ns
spaced sampling points,
corresponding to BCIDs, are
also shown [1]

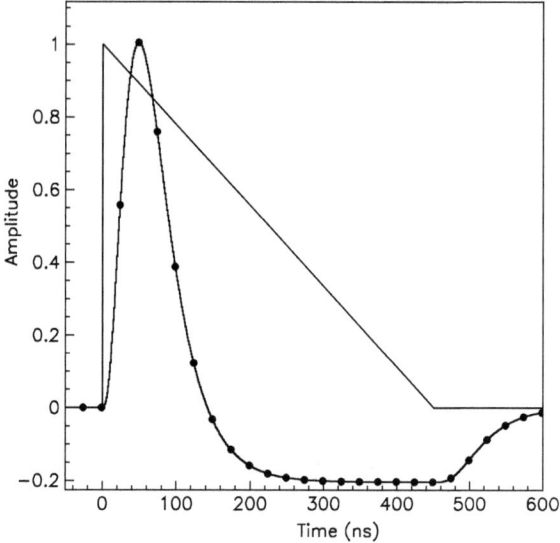

pulse at the triggered BCID is used to dynamically choose the amplifier gain[10] used
for the channel at that given triggered BCID, avoiding saturation but optimising the
signal-to-noise ratio.

In LAr, the triangular pulses undergo bipolar shaping, as shown in Fig. 5.4, giving
them a net integral of zero. The positive part is short and high, followed by a shallow
negative tail of longer duration. It is clear from this figure that the baseline is not
restored until after 600 ns (this example is from the EMB—the exact timing varies
between LAr subdetectors but it is of the same order), which corresponds to 24 bunch
crossings in 25 ns running. The limiting factor here is the drift time, which is not
present in the Tile calorimeter, using a scintillating medium.

The pile-up sensitivity inherent in the slow readout of the LAr calorimeters is
on average compensated for by the bipolar pulse shaping, where the zero integral
on average cancels in-time and out-of-time contributions. The philosophy is that on
average, there will be an equal rate of and signal amplitude from pile-up on top of the
triggering signal as there is activity in the same cells outside triggered events. The
comparatively short duration of the Tile readout allows it to be unipolar, with a pulse
width of 50 ns. A pulse shape quality factor can be used to discriminate between
signals affected and unaffected by pile-up from neighbouring BCIDs.

In general, the signal from an interacting high-energy particle is not contained
within a single cell, as the showers spread both laterally and longitudinally in the
calorimeter.

---

[10]There are discrete gain levels: low, medium, high for LAr; low and high for Tile.

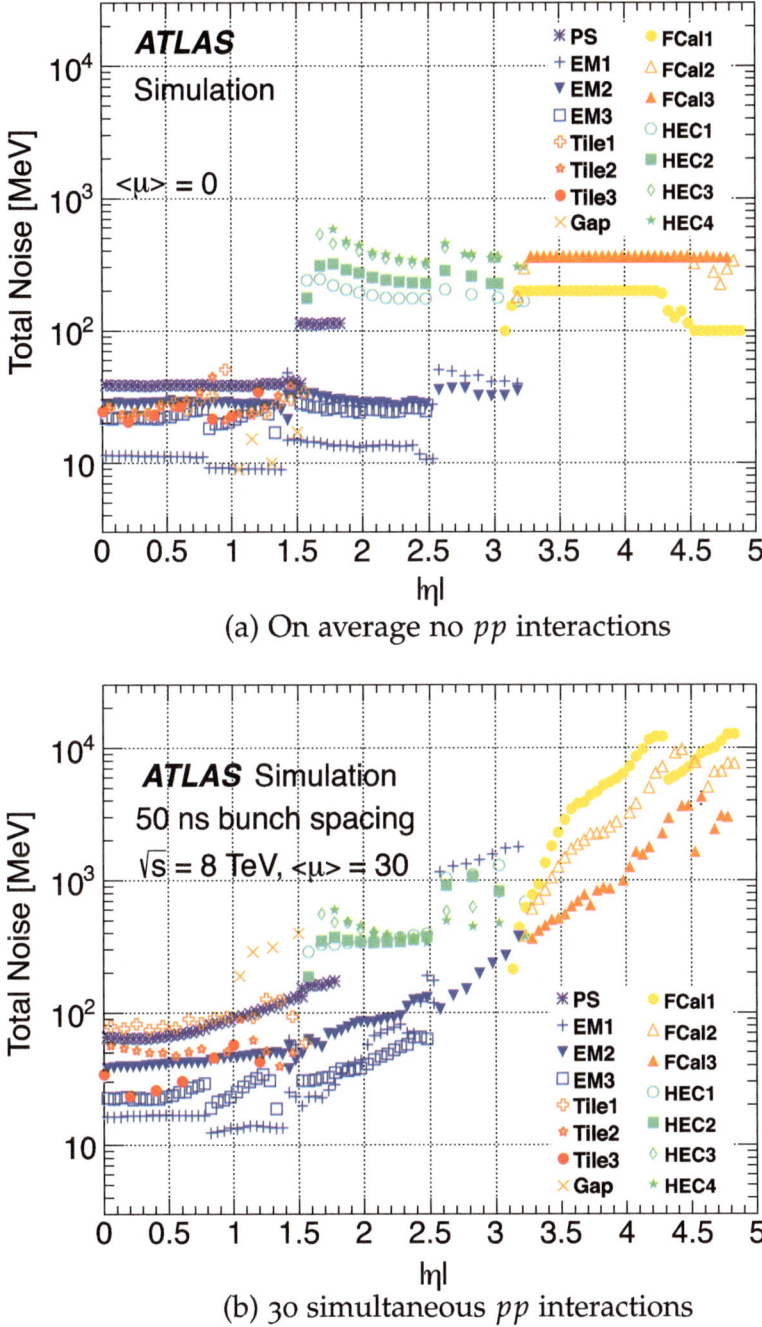

(a) On average no *pp* interactions

(b) 30 simultaneous *pp* interactions

**Fig. 5.5** The average equivalent energy cell noise shown for all the layers in the calorimeter subdetectors, for the case of **a** an average number of interactions of 0 and **b** an average number of 30 simultaneous interactions

**Fig. 5.6** Illustration of the clustering algorithm, restricted to 2D [3]

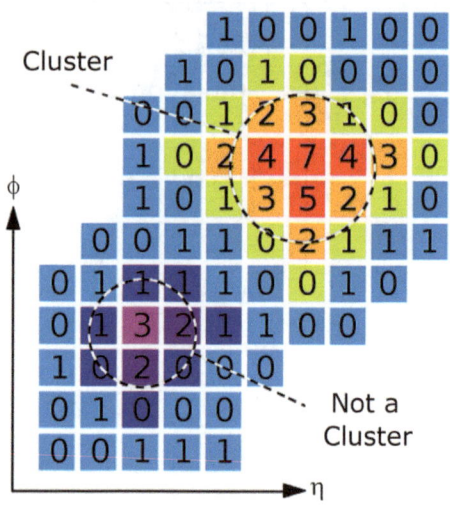

## 5.5  Noise

In every detector, there will be noise. Knowing the noise distribution is the first step towards setting up criteria for identifying signal. Figure 5.5 shows a simulation of the equivalent energy noise levels in the calorimeter under the conditions of 2012 data taking, with 4 TeV beam energy and 50 ns bunch spacing. Figure 5.5a shows the electronic noise, when there is no collision signal in the detector. There is a visible subdetector dependence,[11] where the noise is the largest in the HEC and FCal detectors and smallest in the innermost layer of the EMB. The spread between detectors is also quite wide, about a factor 50. Figure 5.5b shows the noise when there is signal from multiple simultaneous proton collisions overlaid in the detectors. We see that the overlaid collisions contribute large fluctuations in the signal read out from the detector, as expected from having a large particle activity. Here a trend in $\eta$ also emerges that is the most pronounced in the more forward region.

As mentioned in Chap. 3, the LHC operates in a mode where there are many simultaneous interactions, while most likely only one actually triggered the recording of the event. In that sense, the additional interactions become noise.[12]

---

[11] Note the logarithmic vertical scale.

[12] The impact of overlaid interactions, which increases with integration time, is also well described by the established signal processing term *parallel noise*.

## 5.6 Topoclustering

The energy deposits in the calorimeter make *topoclusters* of contiguous cells, based on topological criteria, illustrated in Fig. 5.6. A topocluster is seeded by an energy deposit in a cell which is significantly above the noise level: in the so-called 420 scheme used, it needs to be at least larger than 4 standard deviations ($\sigma$) of the (electronic+pile-up) noise distribution. Once seeded, the topocluster grows by the addition of all neighbouring cells (in three dimensions) with a significance of at least $2\sigma$. Finally all nearest neighbours are added (equivalent to a threshold of $0\sigma$) [2].

As also illustrated in Fig. 5.6, topoclustering introduces noise suppression, without setting a fixed threshold on the energy depositions.

## 5.7 Electromagnetic and Hadronic Scale

The calibrated calorimeter energy deposits clustered into topoclusters are used as input for identifying different physics objects. By this, we mean something that corresponds to measuring the interactions of theoretically defined objects, such as leptons, photons and other particles, with the detector. This is where the reconstruction of the processes in the event begins. But as mentioned before, the ATLAS hadronic calorimeters are non-compensating, meaning that the hadronic and electromagnetic energy scale—the measured energy of, say, a neutron and an electron carrying the same initial energy—will not be the same: the calorimeter *response* will be different. This can be compensated for on the topoclusters, using cluster shapes as a proxy for the shower evolution. Based on depth and location in the calorimeters, lateral shapes, etc., enough discrimination is achieved to assign clusters a hadronic or EM character, and calibrate them accordingly, using MC simulated and test-beam data.

There are two cluster calibration schemes used in ATLAS: the EM and the Local Cluster Weighting (LCW) schemes. For the first, the EM component of energy deposition is used,[13] while in the LCW scheme, the lost energy in clusters classified as hadronic is compensated. In earlier iterations, even though the LCW energy scale was more correct compared to the true deposited energy—the energy response was better—, the energy *resolution* was worse. Since collision data and MC simulations are treated the same way, comparing data and MC on the EM scale will give an apples-to-apples comparison,[14] even if the absolute scale is slightly off.[15] Many analyses have thus chosen to work on the EM scale. Recently, however, jet calibration meth-

---

[13]Remember: hadronic interactions also lead to EM depositions.

[14]This obviously relies on having a proper modelling of the particle composition in the shower! Thankfully, there exist decades worth of data on shower evolution and energy response in different materials and detectors, making the detector simulation remarkably robust. ATLAS has itself been contributing here by sharing test-beam and simulation results with the GEANT4 collaboration.

[15]The scale can also be corrected back to the "true" particle level using the knowledge of the response from MC simulation.

ods taking differences in the hadronisation of quarks and gluons into account[16] have improved the energy resolution of the LCW scale jets, making this a better choice in terms of response *and* equal in terms of resolution.

# References

1. G. Aad et al., The ATLAS experiment at the CERN Large hadron collider. J. Instrum. **3**, S08003 (2008)
2. W. Lampl et al., Calorimeter clustering algorithms: description and performance. ATL-LARG-PUB-2008-002, ATL-COM-LARG-2008-003 (2008)
3. J.J. Goodson, Search for supersymmetry in states with large missing transverse momentum and three leptons including a Z-Boson. Presented 17 Apr 2012. Ph.D. thesis. Stony Brook University (2012)

---

[16]In particular, the different ratio of charged to neutral components, is measured using charged particle tracks.

# Chapter 6
# Jet Finding

In order to reconstruct what we theoretically mean by a jet—a collimated spray of particles of partonic[1] origin—we need a means to localise this spray in our detector. The procedure is to look for localised higher density of energy deposits. In an event display, visualising energy deposits in $\eta - \varphi$ space, our brains have no problem identifying the general regions of interest.[2] For the analysis of a very large number of events, visual inspection is not feasible (not to mention, probably a bit too arbitrary). What is currently used are criterion based algorithms. The requirements on these are that they should reproduce the calculable results in QCD; more precisely, they need to be *infra-red (IR) and collinear safe*. IR safety means that a an additional soft emission cannot change the conclusions of the algorithm. Collinear safety means that a small-angle splitting cannot change the conclusions of the algorithm. Both processes are ubiquitous in QCD—ensuring these properties is a big deal! As an example, measuring the $p_T$ spectrum of leading particles, or using a hard lower energy threshold in the jet finding algorithm tends to be IR-unsafe: soft emissions shift the spectrum and move jets below threshold. Similarly, a fixed "cone" size in $\eta - \varphi$ space tends to be collinear unsafe. There is however a family of algorithms, the sequential recombination algorithms [1], that is nowadays completely dominant precisely because of its IR- and collinear safety. These algorithms will be briefly outlined in the following.

The reader is warned that we will now abandon thinking about jets purely from the perspective of a partonic hadron shower originating from a hard scatter. From

---

[1] $\tau$ leptons are heavy enough to also decay hadronically, producing jets. We will not consider these further as this process is much less likely than jets produced in QCD.

[2] This is a form of pattern recognition, that it would be interesting to pursue in for instance the realm of computer vision.

© Springer International Publishing AG 2017
L.K. Bryngemark, *Search for New Phenomena in Dijet Angular Distributions at √s = 8 and 13 TeV*, Springer Theses, DOI 10.1007/978-3-319-67346-2_6

here on, we use an instrumental definition of jets: a jet is the collection of objects which is the outcome of a jet finding algorithm. Where a connection to the more parton-oriented definition is needed, we will refer to *hard-scatter jets*.

## 6.1  Sequential Recombination Algorithms

For a pair of objects $i \neq j$ with transverse momenta $p_{Ti}$ and $p_{Tj}$, a set of relative distance measures is defined as

$$d_{ij} = \min(p_{Ti}^{2p}, p_{Tj}^{2p}) \frac{\Delta R_{ij}^2}{R^2}, \tag{6.1}$$

$$d_{iB} = p_{Ti}^{2p}, \tag{6.2}$$

where $R$ is the distance parameter of the algorithm, $\Delta R_{ij}^2 = (\Delta \eta_{ij})^2 + (\Delta \varphi_{ij})^2$ and the choice of $p = -1, 0, 1$ defines the $k_t$ [1, 2], Cambridge/Aachen [3, 4] and anti-$k_t$ [5] algorithms, respectively. The squares ensure all distance measures are positive. The subscript $B$ stands for Beam but should not really be interpreted in the proton beam sense.

The algorithm steps are:

1. for all the possible pairs $(i, j)$, compute $d_{ij}$ and $d_{iB}$
2. find the minimum of these
3. if the minimum relative distance is a $d_{ij}$, combine the four-vectors of $i$ and $j$ into a proto-jet
4. if the minimum relative distance is a $d_{iB}$, the four-vector of $i$ becomes a jet and is not considered further
5. repeat until there are no more objects $j$ for which $d_{ij} < d_{iB}$.

Once this procedure is done, the event has been clustered into a set of jets, all having four-vectors, from which the jet $p_T$, direction, $E$ etc can be calculated. In particular, using the relation

$$m^2 = E^2 - p^2, \tag{6.3}$$

the jet mass $m$ can be calculated.

The choice of $p$ as positive, negative or zero means that the object $p_T$ enters as a numerator ($k_t$), denominator (anti-$k_t$), or not at all (Cambridge/Aachen). The distance parameter regulates how far from the jet axis clusters can be considered for addition: if $\Delta R_{ij} > R$, the ratio is larger than unity, which makes it less likely that $d_{ij} < d_{iB}$. Keep in mind that the proto-jet axis is recalculated for every addition, and that this will be dominated by the high-$p_T$ region of the proto-jet. Depending on how much of the proto-jet $p_T$ was added in the previous step, this axis may wander, meaning that the total angular reach of a jet algorithm may exceed $R$.

### 6.1.1  $k_t$

With $p = 1 > 0$, the smallest $p_T$ object will be the starting point for jet finding. This also means, that the jet axis may start to migrate as higher $p_T$ objects are added into the proto-jet.

### 6.1.2  Cambridge/Aachen

With $p = 0$, all $p_T$ measures, in particular $d_{iB}$, are unity. Thus, finding the angular separation is all there is to this algorithm. Again, the jet axis is still sensitive to the location of the high-$p_T$ depositions within the jet, and can migrate.

### 6.1.3  Anti-$k_t$

With $p = -1 < 0$, the highest $p_T$ object will be the starting point for jet finding. This also means, that the jet axis is largely fixed from the start, and the distance parameter $R$ largely governs the angular reach of a jet. Since more objects will be added into the highest $p_T$ proto-jet, clustering continues to add all possible constituents within $R$, before it continues to the highest $p_T$ object outside this jet. Thus, in a dense environment, any locally hardest jet will have circular boundaries,[3] while the ones of lower-$p_T$ jets close by will be crescent shaped.

## 6.2  Jet Catchment Areas

Having thus defined a set of IR- and collinear safe algorithms, we proceed to the concept of the jet catchment area [6]. As pointed out in the original paper introducing the concept, a jet consists—at least theoretically—of point-like particles, making an area ill-defined.

The picture is the following: the jet area is the region in $\eta - \varphi$ space into which a very soft particle would need to fall in order to become clustered with that particular jet. Hence the "catchment" area. This region can be probed by placing *ghosts* of infinitesimal momentum placed at different points in $\eta - \varphi$ and noting to which jet they are clustered. With an IR safe algorithm, the jet finding isn't affected by the addition of a ghost particle, or even by a larger number of them, as long as their momenta are infinitesimal. One can thus lay out a dense grid of ghosts in $\eta - \varphi$ without affecting the jet finding.

---

[3]Exceptions exist, for instance where there are two objects within $R$ that are of comparable $p_T$.

There are thus two imaginable cases for calculating the jet area: one where the jet areas are probed with a single ghost, and one where a dense grid of ghosts is laid out before jet finding starts, participating in the clustering on equal footing (apart from the very different $p_T$) with the four-momenta of energy depositions from the event. This gives rise to the notion of *passive* and *active* areas, which will be defined more precisely below.

## 6.2.1  Active Area

With an event producing a set of particles to be clustered into jets, and a specific set of ghosts $\{g_i\}$ with a density in $\eta - \varphi$ of $v_g$, we can express the scalar area of jet $j$ as

$$A_j = \frac{N_g}{v_g} \qquad (6.4)$$

where $N_g$ denotes the number of ghosts associated to the jet. Since the ghost distribution in $\eta - \varphi$ is random, the area boundaries may differ between different ghost sets. In its full definition, the area is taken as the limit when $v_g \to \infty$, averaged over a large number of ghost set iterations; in practice, $\mathcal{O}(1)$ iterations give precise enough results in relation to the computational cost. Allowing the ghosts to participate in the clustering means that they can cluster among themselves, making pure ghost jets with infinitesimal momentum.

The features of the resulting *scalar* active areas for the three different algorithms described above are exemplified in Fig. 6.1. The different jets are drawn in different shaded regions, and the $p_T$ of its constituents are reflected in the height of the bars, quantified on the vertical axis. The figure shows the irregular shapes of the $k_t$ and Cambridge/Aachen jets, and the circular and crescent shaped anti-$k_t$ jets (bottom). Furthermore it shows the low-$p_T$ dominance in the $k_t$ case (top) compared to the $p_T$ agnostic Cambridge/Aachen (middle): the highest $p_T$ jets (red, green, dark blue) are almost invaded by low-$p_T$ neighbours (magenta, grey) in the $k_t$ case.

If one instead uses the ghost four-momenta $\{g_{\mu,i}\}$ one can define a *four-vector area* as

$$A_{\mu,j} = \frac{1}{v_g \langle g_t \rangle} \sum_{g_i \in j} g_{\mu,i}, \qquad (6.5)$$

where $\langle g_t \rangle$ is the average ghost transverse momentum component, and again, the randomness of the ghost distribution is to be taken properly into account. The transverse component of $A_{\mu,j}$ tends to the scalar area $A_j$ for small $R$.[4]

---

[4]Small $R$ is $R \lesssim 0.4$.

**Fig. 6.1** The jet scalar
active areas found in the
same simulated event,
clustered with (top) $k_t$,
(middle) Cambridge/Aachen
and (bottom) anti-$k_t$. The
vertical axis shows the jet
$p_T$, and pure ghost jets are
omitted for clarity. The same
distance parameter $R = 1.0$
is used for all clustering
algorithms [5]

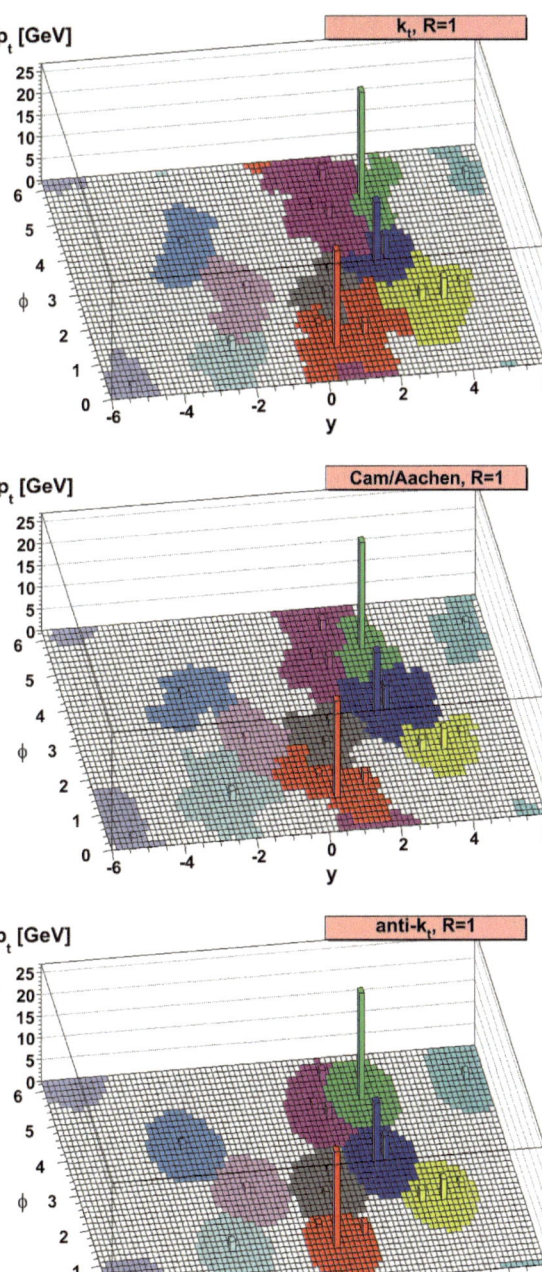

### 6.2.2   Passive Area: The Voronoi Area

As we have seen, the $k_t$ algorithm is very sensitive to soft radiation, and will cluster this first. Thus, when a single ghost is added, it will be associated to one of the particles making up a jet $j$ before any of the pair-wise particle clustering happens. This means that for $k_t$, we can probe the passive area of each non-ghost jet *constituent* independently and add the constituent areas to make up the jet area. It also implies, that the computational cost can be reduced by assigning the geometrical construct of the *Voronoi cell*.

Defined in words, every point inside a constituent's Voronoi cell is closer to that constituent than to any other constituent in the event.[5] The boundaries of the event's Voronoi cells thus indicate the transition points where a ghost would get clustered together with one constituent rather than another. For $k_t$, the distance parameter also enters in the shape of a radius, and the constituent's passive Voronoi area becomes the intersection of the two:

$$A_i^V = V_i \cap C_{i,R}, \qquad (6.7)$$

where $V_i$ is the Voronoi cell of constituent $i$ and $C_{i,R}$ is the circle around it given by the distance parameter $R$. Then the jet Voronoi area is simply

$$A_j^V = \sum_{i \in j} A_i^V. \qquad (6.8)$$

An illustration is given in Fig. 6.2.

The great virtue of the Voronoi area is its much shorter computation time; in the limit of a sparse environment, it is however not a very precise approximation of the active area of a jet, as can be seen from comparing the top figure of Fig. 6.1 to Fig. 6.2.

## 6.3   Jets in ATLAS

ATLAS uses charged tracks or positive massless topoclusters at the EM or LCW scale as jet finding input, forming the jet constituents. The jet clustering framework FASTJET [7] is used to find the jets and calculate their active area, using one set of

---

[5]Mathematically, the definition of a Voronoi cell $V_k$ associated with point $P_k$ is expressed through the points $x$ in some space $X$ as

$$V_k = \{x \in X | d(x, P_k) \le d(x, P_j), \forall j \ne k\}, \qquad (6.6)$$

where $k, j$ are indices ordering the points to which we are assigning Voronoi cells. In our application, $P_{k,j}$ is a jet constituent while the $x$ locations would be probed by a ghost.

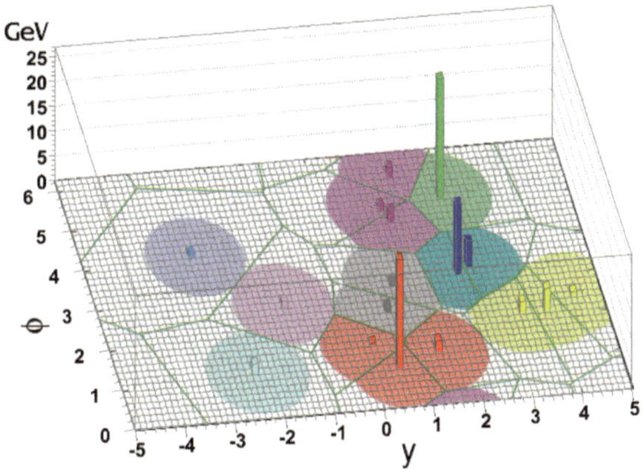

**Fig. 6.2** The jet Voronoi areas in the same simulated event as in Fig. 6.1, calculated using $R = 1$, indicated in shaded regions. The green lines are the boundaries of the Voronoi cells. The vertical axis shows the jet $p_T$ [6]

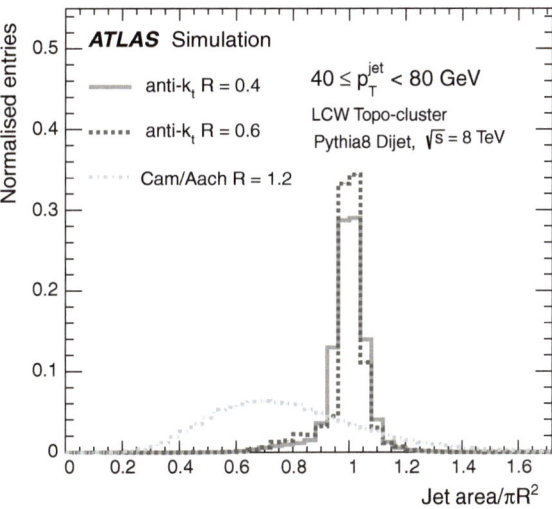

**Fig. 6.3** The area in units of $\pi R^2$ for anti-$k_t$ jets with $R = 0.4, 0.6$ and Cambridge/Aachen jets with $R = 1.2$. The circular tendency of anti-$k_t$ centres those distributions around 1, while the Cambridge/Aachen jets have a large spread

ghosts.[6] This is done for a range of *jet definitions*—combinations of jet algorithms and distance parameters—but the ones where baseline calibration is provided are anti-$k_t$ jets with $R = 0.4, 0.6$.

Fig. 6.3 shows the area distributions expressed in units of $\pi R^2$ for three jet definitions: two choices of $R$ for anti-$k_t$, and one for Cambridge/Aachen. For circular jets, the distribution should be centred around 1. This is the case for anti-$k_t$, with a long tail to smaller areas from crescent-shaped jets. The irregular area for Cambridge/Aachen jets is seen here in terms of a very wide distribution.

Apart from making up pure track jets, individual tracks can also be matched to jets using a *ghost association* scheme [8]. Here the track momenta are temporarily

---

[6]Computation time would scale at best linearly with number of iterations, while the resolution improves roughly with the square root of that number.

set infinitesimally small,[7] and the association of tracks to a jet is made using the same techniques as for active area calculation. This matching scheme has an advantage over cone-based association ("$\Delta R$" matching from the jet axis) when the jet shape is non-circular. The matched tracks can be used for instance to improve jet calibration or selection and for flavour tagging.

## 6.3.1  Jet Calibration

Since jets are complicated composite objects,[8] jet calibration is a procedure involving a sequence of steps. It will only be briefly explained here, and the interested reader is referred to Ref. [9] and references therein. The result of the calibration is that the jet is brought to the *Jet Energy Scale* (JES).

The first step is the topocluster energy calibration mentioned in Sect. 5.7. Then jet finding results in a set of jets at the EM or LCW constituent energy scale. An origin correction makes the jets point back to the primary vertex rather than the detector origin. The jet four-vectors are then corrected for pile-up effects. This is a step where I have contributed much work, and it will be discussed in some depth in Chap. 7. After pile-up correction, a MC based energy scale correction is applied, which brings the jets to particle-level energy. Since MC modelling can't be trusted to provide a perfect representation of data, this correction has been validated using *in-situ* techniques where the calibrated jet $p_T$ is compared to the $p_T$ of a balancing reference object. The reference objects can be photons, $Z$ bosons, or other jets, and are selected using criteria ensuring that a $p_T$ balance is indeed expected in these events. This step will affect both the energy and the direction of the jet, i.e., the full jet four-vector. Once the scale is corrected, a correction improving resolution is applied: it uses tracking to adjust for flavour dependent effects in the energy measurement.[9] In addition, it corrects for energy leakage when a jet "punches through" the calorimeter and deposits energy in the muon spectrometer. This correction is based on the number of muon segments behind a calorimeter jet. In practice, punch-through is a rare phenomenon, but it is enhanced for the highest energy jets, in the central regions of the calorimeter where the detector material is the thinnest in terms of radiation and absorption lengths. Finally, a residual $\eta$ dependent correction, the $\eta$ *intercalibration* [10], is applied to data only, using dijet balance to ensure a uniform energy response between different regions of the detector. The relative uncertainty associated with the jet calibration is summarised in Fig. 6.4.

The uncertainty has been derived using the aforementioned *in-situ* methods. As one would expect from the general properties of calorimeters, the relative uncertainty

---

[7]"infinitesimally" small: $p_T = 1$ eV.

[8]All experimental objects are composite, as they are reconstructed from a large number of signals. The energy calibration of an object is thus a bit more involved than the calibration of a single detector signal.

[9]Remember the different response for charged and neutral hadrons.

**Fig. 6.4** The relative
uncertainty on the jet energy
scale shown as function of
(**a**) jet $p_T$ for central jets and
(**b**) jet $\eta$ for jets at
$p_T = 40$ GeV [9]

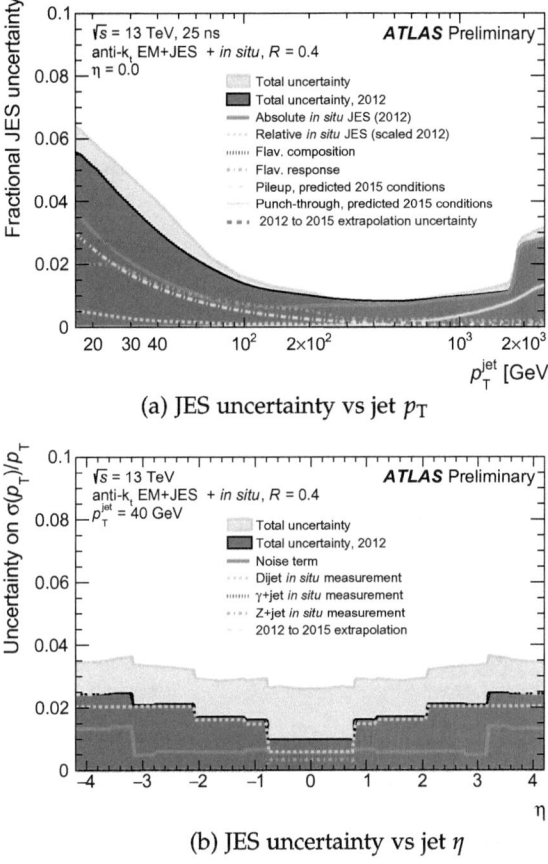

(a) JES uncertainty vs jet $p_T$

(b) JES uncertainty vs jet $\eta$

decreases with jet $p_T$, as seen in Fig. 6.4a. However, at $\sim 2$ TeV it rises sharply and
becomes flat in $p_T$. The increased uncertainty is due to large statistical uncertainties
in the *in-situ* methods. The uncertainty also varies with $\eta$, as shown in Fig. 6.4b. Note
that the uncertainty shown here is for jets at $p_T = 40$ GeV, which has comparatively
large relative uncertainty.

## 6.3.2  Jet Cleaning

Once we have a set of calibrated jets, jet *cleaning* [11] is applied to ensure that
"fake" jets originating from for instance beam-induced or cosmic ray backgrounds,
or calorimeter noise bursts, are not included. The selection criteria are designed to
capture these rather diverse signatures and are based on, among others, the fraction

of the jet energy deposited in the HEC, calorimeter pulse quality factors, the fraction of EM energy, and the amount of negative energy in the cells of the jet.

While it is individual jets that are induced by non-collision sources, to avoid introducing a bias, cleaning is applied as a rejection of whole events. An event is discarded if any of the two highest $p_T$ jets are deemed fake, or any other jet carrying more than 30% of the second highest $p_T$ is. It was established that below this fraction, the impact from non-collision sources would not introduce changes in the ordering of jets.

# References

1. S.D. Ellis, D.E. Soper, Successive combination jet algorithm for hadron collisions. Phys. Rev. D **48**, 3160–3166 (1993)
2. S. Catani et al., Longitudinally invariant Kt clustering algorithms for hadron hadron collisions. Nucl. Phys. B **406**, 187–224 (1993)
3. Y.L. Dokshitzer et al., Better jet clustering algorithms. J. High Energy Phys. **08**, 001 (1997)
4. M. Wobisch, T. Wengler, Hadronization corrections to jet cross-sections in deep inelastic scattering. in *Monte Carlo generators for HERA physics. Proceedings, Workshop, Hamburg, Germany, 1998–1999* (1998)
5. M. Cacciari, G.P. Salam, G. Soyez, The Antik$_t$ jet clustering algorithm. J. High Energy Phys. **04**, 063 (2008)
6. M. Cacciari, G.P. Salam, G. Soyez, The catchment area of jets. J. High Energy Phys. **04**, 005 (2008)
7. M. Cacciari, G.P. Salam, G. Soyez, FastJet user manual. Eur. Phys. J. C **72**, 1896 (2012)
8. ATLAS Collaboration, Performance of jet substructure techniques for large-R jets in proton-proton collisions at $\sqrt{s}$ = 7 TeV using the ATLAS detector. J. High Energy Phys. **09**, 076 (2013)
9. ATLAS Collaboration, Jet calibration and systematic uncertainties for jets reconstructed in the ATLAS detector at $\sqrt{s}$ = 13 TeV. ATLPHYS- PUB-2015-015 (2015)
10. ATLAS Collaboration, In-situ determination of the ATLAS forward jet energy scale and uncertainty using dijet events at $\sqrt{s}$ = 13 TeV. ATLAS-COM-CONF-2015-023 (2015)
11. ATLAS Collaboration, Selection of jets produced in 13 TeV protonproton collisions with the ATLAS detector. ATLAS-CONF-2015-029 (2015)

# Chapter 7
# Pile-Up in Jets

The work described in this chapter is also partly documented in Refs. [1, 2]. It describes the adaptation of a pile-up subtraction technique for jets, proposed and validated in the absence of a detector [3], for use within ATLAS. The work was largely done during 2012, using simulated 7 and 8 TeV collisions, and the method is the new standard jet pile-up correction used in ATLAS since the end of that year.

## 7.1 Pile-Up Observables

As described in Sect. 5.4, the calorimeter signal readout is longer than one bunch crossing. The shaped LAr signal is 450–600 ns long, corresponding to 18–24 bunch crossings at 25 ns spacing. It was concluded that this makes in particular the LAr calorimeter pulses sensitive not only to overlaid signal from other interactions in the same BCID, *in-time pile-up*, but from neighbouring bunch crossings as well, and in fact to activity in the fairly distant history, *out-of-time pile-up*.

Although the bipolar shaping is designed to on average cancel the in-time and out-of-time pile-up contributions to the signal, in reality there will be large fluctuations leading to imperfect cancellation. In order to quantify the impact, we introduce two quantities describing the number of pile-up interactions in an event: the *number of reconstructed primary vertices* $N_{PV}$, and the *number of simultaneous inelastic collisions*, $\mu$:

$$\mu = \frac{L_0 \sigma_{inelastic}}{n_c f_{rev}}, \tag{7.1}$$

where $L_0$ is an instantaneous luminosity, $\sigma_{inelastic}$ is the cross section for inelastic $pp$ interactions, $n_c$ is the number of colliding bunch pairs and $f_{rev} = 11.245$ kHz is the revolution frequency of the LHC.[1] As we have seen, in practice instantaneous

---

[1] Note the resemblance to Eq. 3.5.

© Springer International Publishing AG 2017
L.K. Bryngemark, *Search for New Phenomena in Dijet Angular Distributions at $\sqrt{s} = 8$ and 13 TeV*, Springer Theses, DOI 10.1007/978-3-319-67346-2_7

bunch crossing luminosity is not measured, but averaged over a lumiblock. We are thus restricted to using the average, $\langle\mu\rangle$. The actual number of interactions in an event follows a Poisson distribution with mean $\langle\mu\rangle$, measured for the corresponding lumiblock.

In contrast, $N_{PV}$ is an event-by-event quality. A primary vertex is defined as a vertex reconstructed from at least two tracks with $p_T > 400\,\text{MeV}$, consistent with the LHC beam spot. The hard-scatter primary vertex, the one considered to be associated to the interaction triggering the event, is defined as the primary vertex with the highest $\sum_{tracks}(p_T^2)$.

These two quantities are highly correlated, with the average relation in 2012 data of $N_{PV} \approx 0.5\mu$. But there are large fluctuations:

- the number of interactions is not known for the individual event, only $\langle\mu\rangle$ for the lumiblock
- even for a well-defined $\mu$, there can be fluctuations in for instance the vertex reconstruction efficiency, leading to fluctuations in $N_{PV}$.

$N_{PV}$ is a measure of the in-time pile-up contributions only, since the tracking detectors are fast and readout is completed before the next bunch crossing. On the other hand, $\langle\mu\rangle$ encompasses both: the number of interactions in the previous bunch crossing belongs to the same Poisson distribution as the number in the current one, with the same $\langle\mu\rangle$, and we can thus expect about the same level of pile-up in the non-triggered bunch crossing as in the triggered one. Using both observables, they can be decorrelated to isolate the impact of in-time and out-of-time pile-up, by fixing one while letting the other vary.

### 7.1.1  Impact of Pile-Up on Jets

The impact on jets from pile-up can be divided into three categories:

- response: energy is added to or subtracted from the measured hard-scatter jet energy,
- resolution: the jet energy measurement is smeared,
- multiplicity: additional jets from pile-up are reconstructed in the event.

This defines the measures we'll use in the following to quantify the impact of pile-up on jets:

- the slope of the jet $p_T$ or mass response with respect to $N_{PV}$ or $\langle\mu\rangle$: $\partial p_T/\partial N_{PV}$ and $\partial p_T/\partial\langle\mu\rangle$, and correspondingly for mass $m$;
- the jet $p_T$ or mass resolution expressed as the RMS of $p_T^{reco} - p_T^{true}$ or $m^{reco} - m^{true}$;
- the number of jets in an event above a certain $p_T$ threshold.

The previous pile-up correction method in ATLAS used a parameterisation in $N_{PV}$ and $\langle\mu\rangle$, derived from the dependence on these observed in MC, to correct for pile-up in jets. That method was an average correction, restoring the average jet response to be independent of pile-up. In the following we shall outline the current method, which measures the amount of pile-up on an event-by-event level, thus improving also the resolution in the presence of pile-up.

## 7.2  Jet-Area Based Correction

The current method implemented in ATLAS is based on the area concepts discussed in Sect. 6.2. The idea is, that the jet area is a measure of how susceptible a given jet is to pile-up, as it reflects the extent in $\eta - \varphi$ space within which a soft particle would be associated to the jet. IR and collinear safety ensures that we can measure that susceptibility without worrying that the $p_T$ spectrum and distribution in $\eta - \varphi$ of the pile-up would distort our identification of the jet boundaries. It is assumed here that pile-up is diffuse, soft radiation.

The other ingredient is measuring the pile-up $p_T$ density in the given event. Consider an event with a few hard-scatter jets and some number of overlaid interactions, contributing diffuse radiation across the $\eta - \varphi$ cylinder of the detector. Once the event is clustered with a jet finding algorithm, this activity will be contained in a set of jets with a given $p_T$ and area. These are used to calculate a median $p_T$ density $\rho$:

$$\rho = \text{median}\left(\frac{p_T^j}{A_j}\right). \tag{7.2}$$

The median is taken to avoid a large influence from the hard-scatter jets, which will contribute a high-density tail to the distribution.

If for a given jet, the area is multiplied with the $p_T$ density, one obtains a measure of how much of that jet's $p_T$ was contributed by pile-up. The jet-area based correction amounts to precisely this subtraction from the jet four-vector. Note that the individual jet constituents remain the same: only the final four-vector is corrected.

### 7.2.1  The $\rho$ Calculation: Algorithm Choices

There is no need for the clustering for the $\rho$ calculation to return regular jets dominated by the hard scatter—on the contrary, it should be dominated by the low-$p_T$ energy deposits from pile-up. The possibility to use the Voronoi area (introduced in Eq. 6.7) makes $k_t$ an attractive choice, and this is used for the $\rho$ calculation.[2] A distance parameter $R_{kt} = 0.3 - 0.4$ was found to be the optimal compromise between an increased sensitivity to the hard-scatter activity in the limit of few jets, and a large

---

[2]Later, a so-called *grid method* has been implemented in FASTJET, which simply slices the event up in pieces of equal size, making the calculation of $\rho$ even faster. This has not yet been implemented in ATLAS.

**Fig. 7.1** $\rho$ calculated using in a window in $\eta$ of width $\Delta\eta = 0.7$, sliding in steps of $\delta\eta = 0.1$, shown as function of the $\eta$ of the midpoint of the window. Curves for different values of $\langle\mu\rangle$ are shown

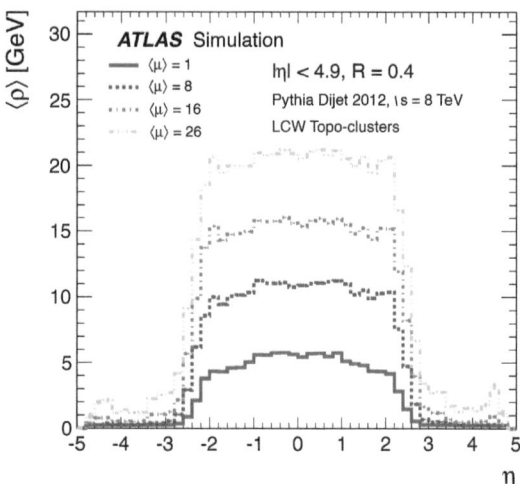

number of very low $p_T$ jets making $\rho$ tend to 0 in the limit of sparse events. For some studies, $R_{kt} = 0.3$ was used, but the final choice was $R_{kt} = 0.4$.

### 7.2.2  The $\rho$ Calculation: $\eta$ Range

Figure 7.1 shows the average $\rho$ in simulated dijet events, calculated using a slice in $\eta$ of width $\Delta\eta = 0.7$, sliding in steps of $\delta\eta = 0.1$ across the full $\eta$ range of the ATLAS detector. Curves for different values of $\langle\mu\rangle$ are overlaid. There is a visible dependence on $\langle\mu\rangle$ in the central region, while the curves all drop outside $|\eta| = 2.0$. At higher $\eta$, $\rho$ is mostly 0, largely independent of $\langle\mu\rangle$.

As we shall see, this behaviour is not caused by an absence of pile-up in the forward region. Instead, it is a matter of granularity. As described in Chap. 5, the ATLAS calorimeters are far from uniform in $\eta$: gaps or other transition points are largely washed out in this sliding window measurement, but the effect of granularity is not. To understand it, we need to consider the noise conditions for topoclustering. In Sect. 5.6, we saw that a signal equivalent to $4\sigma$ of the noise distribution is needed to seed a cluster. This condition is the same across all calorimeters, even if the absolute size in MeV of $4\sigma$ varies, as shown in Fig. 5.4b. But, the probability of registering at least one instance of a large deviation is smaller with few samplings of a given distribution, than with a large number.[3] The fine granularity of the central calorimeters corresponds to a large number of samples, while the segmentation of the forward region is much coarser (cf. Table 5.2). Calculating optimistically, there are 16 times fewer cells per unit area in $\eta - \varphi$ in the forward region than in the central.

---

[3] An analogy from Ariel: compare the probability of finding at least one crying baby at home compared to in a nursery school.

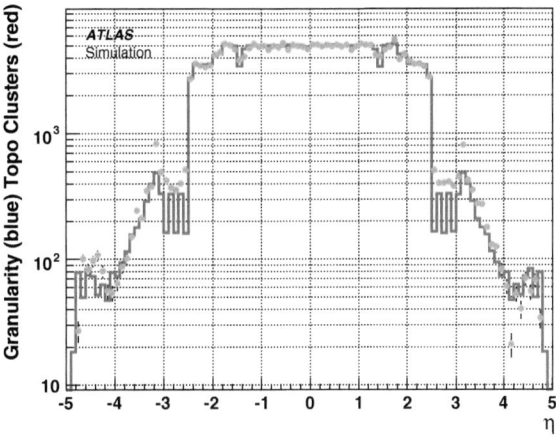

**Fig. 7.2** The profile in $\eta$ of the calorimeter cell granularity, overlaid with the cluster distribution. A close correspondence is seen

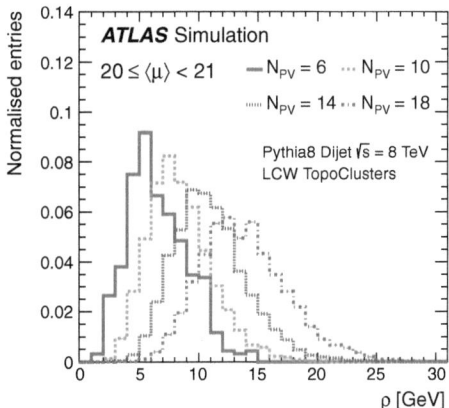

**Fig. 7.3** $\rho$ for different $N_{PV}$, at fixed $\langle\mu\rangle$. The figure shows the fluctuations in the pile-up density captured by $\rho$ but not fully by the $N_{PV}$ and $\langle\mu\rangle$ variables

This reasoning is confirmed in Fig. 7.2, which shows an overlay of the granularity and cluster distributions in $\eta$. The shapes of the two curves follow each other closely, as well as the $\eta$ profile of $\rho$ in Fig. 7.1.

In conclusion: there is very little information to be gained from measuring $\rho$ beyond $|\eta| = 2.0$; rather, there is a risk of diluting the information by measuring mostly empty regions except for when there is hard-scatter activity seeding clusters. The range used for the $\rho$ calculation is thus the entire event within $-2.0 \leq \eta \leq 2.0$.

Figure 7.3 shows example distributions of $\rho$ thus obtained, for different $N_{PV}$ at a fixed $\langle\mu\rangle$. It is clear that even for a given $N_{PV}$ and $\langle\mu\rangle$, there are large fluctuations in the pile-up activity, which are captured in the $\rho$ calculation.

**Fig. 7.4** $p_T^{reco} - p_T^{true}$ vs $N_{PV}$ for a fixed $\langle\mu\rangle$ representative for 2011 data taking, before and after $\rho \cdot A$ subtraction

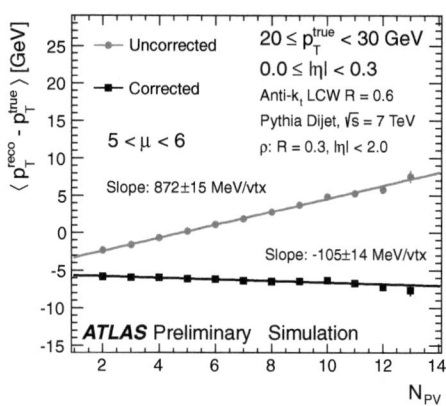

## 7.3   Method Performance

The following section outlines the assessment of how well jets can be corrected for pile-up using the jet-area based method. Even if its event-by-event character includes event-level fluctuations, the only local fluctuations taken into account are those reflected in the individual jet area. This section will describe both the properties of $\rho$ and the impact of the correction on several observables, keeping in mind the main effects listed in Sect. 7.1.1.

### 7.3.1   Response

Using the four-vector area rather than the scalar area, the full jet four-vector can be corrected. This means that not only the jet $p_T$ but also the mass is affected in the correction. The impact on these two variables will be addressed in turn.

#### $p_T$ Response

Most figures in this section are based on graphs like the one shown in Fig. 7.4. It shows the dependence of a jet quantity ($\langle p_T^{reco} - p_T^{true}\rangle$) on a pile-up measure ($N_{PV}$), before and after the $\rho \cdot A$ correction, in some region of phase space (here, a $p_T$ and $|\eta|$ range). It also shows a linear fit to the trend, which captures the dependence very well.

Figure 7.5 is based on this type of graph, and the subfigures show the behaviour in $|\eta|$ for four different jet definitions, keeping the $p_T$ range fixed. The y-axis shows the slopes of the linear fits of $p_T^{reco} - p_T^{true}$ vs $N_{PV}$. $\langle\mu\rangle$ is kept fixed at its average value for 2012 data taking to isolate the impact of in-time pile-up.

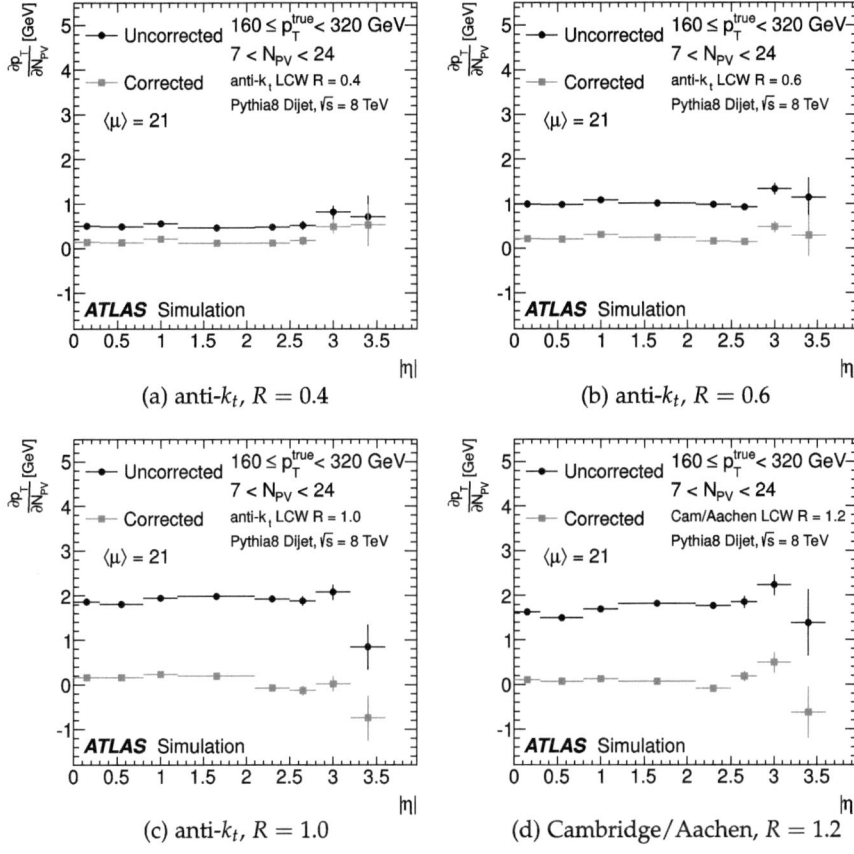

**Fig. 7.5** The dependence of jet $p_T$ on $N_{PV}$ as function of $|\eta|$, before and after $\rho \cdot A$ subtraction for jets clustered with **a** anti-$k_t$, $R = 0.4$, **b** anti-$k_t$, $R = 0.6$, **c** anti-$k_t$, $R = 1.0$, **d** Cambridge/Aachen, $R = 1.2$

The uncorrected slopes, in black, depend strongly on the jet area, which is one of the assumptions of the method.[4] As shown earlier in Fig. 6.5 there is no simple $\pi R^2$ behaviour for Cambridge/Aachen jets. The increased $R$ doesn't automatically give a larger slope in Fig. 7.5d. The corrected slopes, in red, show no area dependence. There is a small residual slope of $\mathcal{O}(100)$ MeV/vertex, of similar size for all jet definitions, and some features in $\eta$ are seen. Most strikingly, the shape in $|\eta|$ before and after correction are the same: the correction merely introduces a downward shift. This is explained by $\rho$ and $A$ both being independent of jet $\eta$.

Here we can make the observation that in the presence of a jet, there is indeed pile-up in the forward region, even if the cluster multiplicity there is mostly independent of $\langle\mu\rangle$. This is explained by the hard-scatter jet seeding the clusters. The in-time

---

[4]The area of a circle with radius $R = 0.4$ is half as large as with $R = 0.6$.

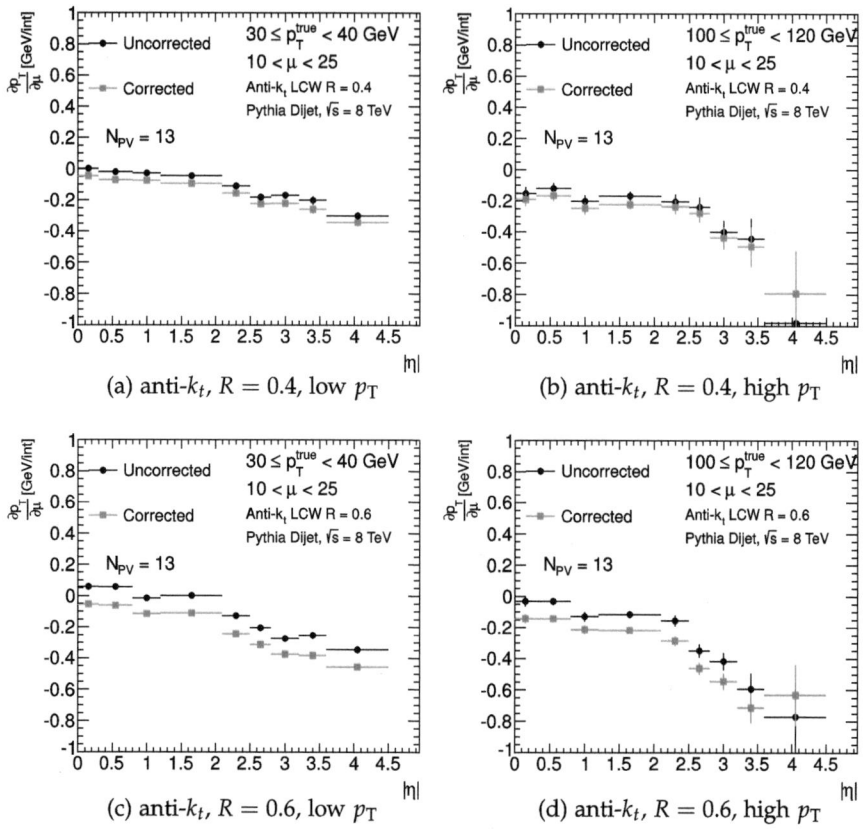

**Fig. 7.6** $\langle\mu\rangle$ dependence of jet $p_T$ vs $|\eta|$, before and after $\rho \cdot A$ subtraction for **a** anti-$k_t$ $R = 0.4$ jets with $30 \leq p_T^{\text{true}} < 40\,\text{GeV}$, **b** anti-$k_t$ $R = 0.4$, $100 \leq p_T^{\text{true}} < 120\,\text{GeV}$, and **c, d** anti-$k_t$ $R = 0.6$ jets in the same $p_T$ intervals. The slopes before correction depend strongly on jet $p_T$, but only in the central region, and weakly, on the jet area

pile-up thus adds energy to clusters that would already have been there. We also see that the $\eta$ dependence of the in-time pile-up is moderate.

Turning to the dependence on out-of-time pile-up, the equivalent $|\eta|$ dependence figures are shown for three jet definitions in Fig. 7.6. $N_{PV}$ is kept fixed at its average for 2012 data taking to isolate the dependence on out-of-time pile-up. Before correction, there is only a small positive slope in the central region, and it quickly drops off at higher $|\eta|$. After correction, the $p_T$ dependence has in most cases become non-zero and negative. This means that a quantity with positive dependence on out-of-time pile-up ($\rho \cdot A$) is subtracted from a quantity with no or negative dependence. As before, there is no $\eta$ dependence in $\rho$, so the correction merely introduces a shift, preserving the shape in $|\eta|$. Figures 7.6a, b show $R = 0.4$ anti-$k_t$ jets at $30 \leq p_T < 40\,\text{GeV}$ and $100 \leq p_T < 120\,\text{GeV}$, respectively. Comparing these two, there is a clear $p_T$ dependence in the uncorrected (black) curves: both an offset across all $|\eta|$, and a

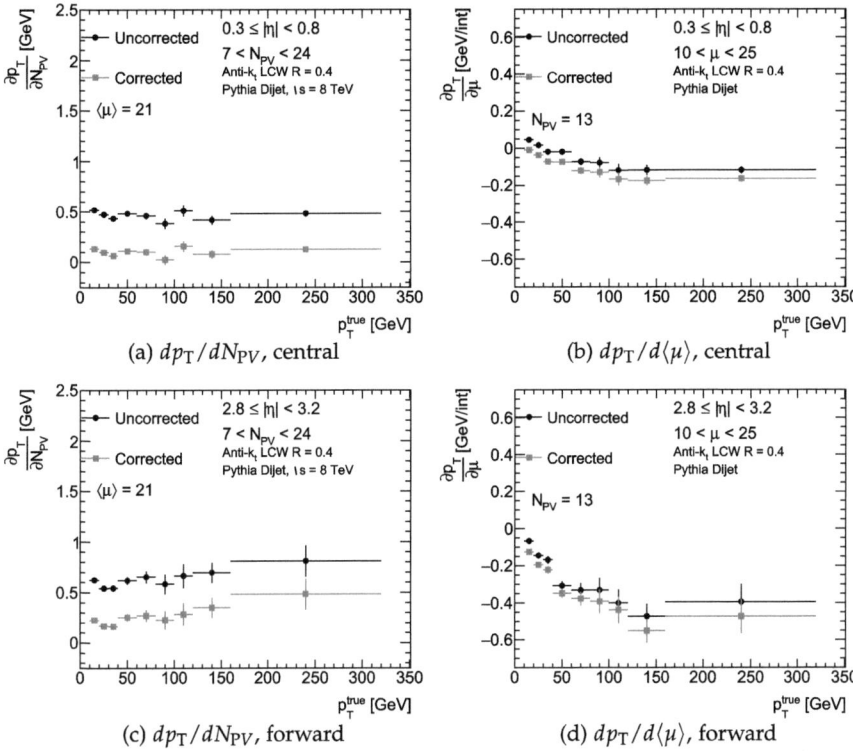

**Fig. 7.7** The dependence of $p_T$ on **a,c** $N_{PV}$ and **b,d** $\langle\mu\rangle$, as function of $p_T$, before and after $\rho \cdot A$ subtraction, for jets clustered with anti-$k_t$, $R = 0.4$. The top row shows central $|\eta|$ and the bottom more forward

quicker drop-off at higher $|\eta|$ for higher $p_T$. Figures 7.6c,d shows $R = 0.6$ anti-$k_t$ jets in the same two intervals. Comparing the two $R$ choices for the same $p_T$ range, there is not much area dependence of the out-of-time pile-up impact at high $|\eta|$, while there is a little in the central region. The correction of course depends strongly on area, making the shift in the corrected points larger for $R = 0.6$.

The $p_T$ dependence of the impact of both in-time (left) and out-of-time (right) pile-up is shown in Fig. 7.7, for two choices of $|\eta|$: very central (top) and more forward (bottom). The dependence is positive in $N_{PV}$ and mostly negative in $\langle\mu\rangle$, and more pronounced at higher $|\eta|$.

The key to understanding these features is the interplay of granularity, overlap, the LAr bipolar pulse shaping and topoclustering. Looking at $\rho$ is instructive; in particular, its dependence on the in-time and out-of-time cluster occupancy. Figure 7.8a shows the mean and width of fits to the peak of $\rho$ distributions like those shown in Fig. 7.3, while Fig. 7.8b in turn summarises the slopes vs $\langle\mu\rangle$ in Fig. 7.8a as function of $N_{PV}$. The fitted distributions have been binned in $\langle\mu\rangle$ and the true $N_{PV}$, the number of primary vertices from the generator truth record rather than after reconstruction.

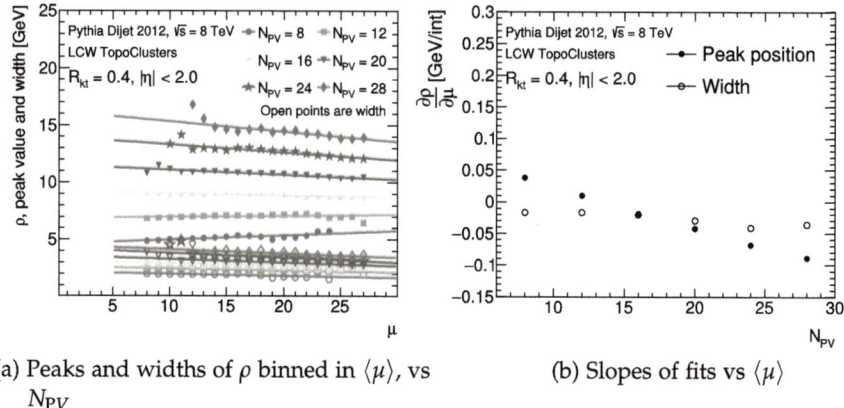

(a) Peaks and widths of $\rho$ binned in $\langle\mu\rangle$, vs $N_{PV}$

(b) Slopes of fits vs $\langle\mu\rangle$

**Fig. 7.8** Mean and width of Gaussian fits to the peak of distributions like in Fig. 7.3, but using the true $N_{PV}$. **a** Linear fits to the evolution in $\langle\mu\rangle$ for a given $N_{PV}$. **b** The slope vs $N_{PV}$ of the linear fits in **a**

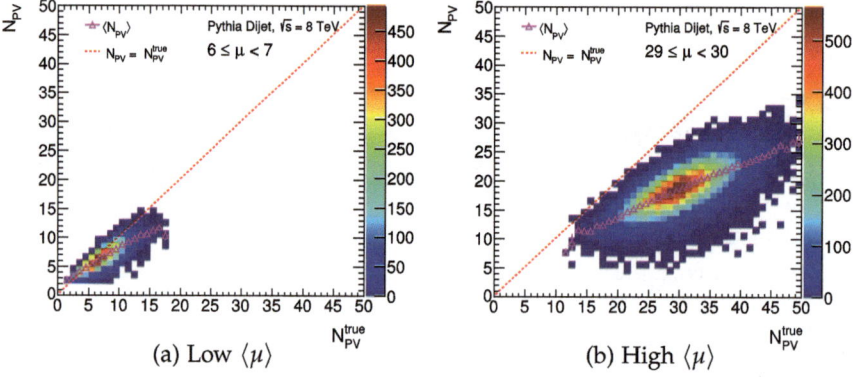

(a) Low $\langle\mu\rangle$

(b) High $\langle\mu\rangle$

**Fig. 7.9** The relationship between the number of reconstructed and true primary vertices, for two regimes of $\langle\mu\rangle$. The reconstruction efficiency is less than unity, and decreases with $N_{PV}^{true}$

This number gives a better account of the in-time occupancy in the events than the reconstructed $N_{PV}$, since the vertex reconstruction efficiency is often less than unity, and non-linear with pile-up activity, as shown in Fig. 7.9.

$\rho$ is calculated from positive clusters in the central region, where there is high granularity and generally small overlap between the clusters. Figure 7.8 shows that when there is small in-time occupancy, $\rho$ has a positive dependence on $\langle\mu\rangle$–meaning, on out-of-time pile-up. This must then come from energy deposited in the very neighbouring bunch crossings, where the LAr pulse shaping produces positive energy contributions. As the in-time occupancy increases, the slope with $\langle\mu\rangle$ becomes smaller and even negative: energy is being *subtracted* by the out-of-time pile-up. Now, a higher in-time occupancy translates to larger overlap between out-of-time pile-up and the in-time pile-up clusters included in the $\rho$ calculation. We thus see that when the overlap increases, the net effect is energy subtraction. The negative energy comes from the negative tail in LAr, which is much longer than the positive part. Overlap with the negative-energy part is thus more probable. In turn, this means that the positive slope in the low-occupancy case *cannot* come from overlap—this causes an overall subtraction of energy—so it must come from isolated positive out-of-time clusters.

To summarise: when an energy deposition overlaps with out-of-time pile-up, this leads to energy subtraction. Out-of-time pile-up can contribute positive energy if there is enough room for it to make isolated positive clusters. Given the $\eta$ dependence of the calorimeter granularity, this immediately introduces an $\eta$ dependence in the impact of out-of-time pile-up. In the central region, it will often be positive. In the forward region, it will only be seen if there is an overlapping seeding positive energy contribution, and as an energy subtraction.

Mapping this to jets, there is in the central region often enough room to allow for non-overlapping out-of-time clusters, especially for large distance parameters. This gives a non-negative or even positive dependence on $\langle\mu\rangle$: cf. Fig. 7.5a,c. Comparing these two, which are the same events clustered with different distance parameters, we see that the negative dependence comes from the core of the jet (making up a larger fraction of the area of narrower jets): there is a positive dependence in the central region only for the larger jets. In the forward region, we have seen that the large cells make the cluster occupancy outside jets mostly zero, and only the overlaps contribute, making the jet out-of-time pile-up dependence negative. Here there is no strong area dependence, since negative clusters are not included in the jet clustering. The area dependence of the positive slope with in-time pile-up in the forward region saturates for larger $R$, as shown for the full $|\eta|$ range in Fig. 7.10. For $R = 1.0$, the positive slope is reduced at higher $|\eta|$. This must then mean that the area dependence for $R = 0.4$–$0.6$ in this region comes from the hard-scatter jet radiation seeding clusters, while as the area increases even more, the pile-up contribution doesn't, as in-time pile-up only rarely contributes additional clusters and mostly contributes positive energy on the fringes of seeded clusters.

In fact, Fig. 7.10 can raise a question mark as to the validity of the anti-$k_t$ area as a measure of the catchment area of the jet in the coarse granularity environment of the forward region, at least in the limit of larger $R$. We will return to this later.

Figure 7.11 shows the distribution of negative and positive clusters in the vicinity of the leading[5] jet, for different $|\eta|$, $p_T^{lead}$ and $\langle\mu\rangle$. First of all, the positive cluster distribution gets narrower at higher $p_T$ and $|\eta|$, making the jet increasingly dominated by the core. This explains the out-of-time pile-up dependence. It is also clear that on the periphery, the cluster density grows with $\langle\mu\rangle$, while at the core it is 0 in the forward region, and small in the central. This is of course inherent in the selection, where $dR = 0$ means that we are on the jet axis, and there is no room for negative clusters since the jet is built from positive clusters, where the jet axis dominated by the highest $p_T$ depositions.

---

[5]*Leading*: carries the highest $p_T$.

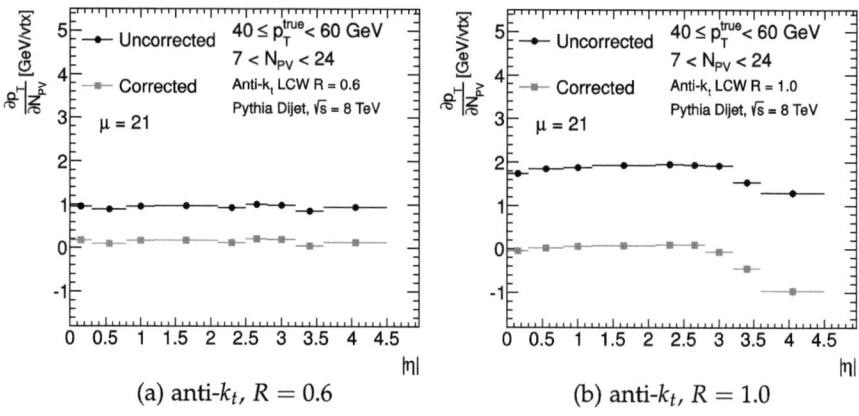

**Fig. 7.10** The dependence of $p_T$ on $N_{PV}$, as function of $|\eta|$ before and after $\rho \cdot A$ subtraction for jets at $p_T^{true} = 40$–$60$ GeV, clustered with anti-$k_t$ using **a** $R = 0.6$ and **b** $R = 1.0$. The slopes before correction depend strongly on jet area in the central region, but only weakly in the forward region

The residual in-time dependence after correction reflects the different sensitivity of $\rho$ and jets, coming from the core (independent of area). The over-correction with respect to out-of-time dependence does too: overall it's a small over-correction on top of a large initial negative dependence. Since $\rho$ reacts differently to out-of-time pile-up than jets do, this dependence needs to be dealt with outside the method, and an additional residual correction is used, which uses a parameterisation in $N_{PV}$ and $\langle\mu\rangle$, derived from MC. This is the same method as the one previously used in ATLAS, with the difference that the parameterised correction is smaller after $\rho \cdot A$ correction. It is substantial only in the forward region, where the out-of-time dependence is large. The performance of the residual correction is exemplified in Fig. 7.12, showing that the dependence of jet $p_T$ on both $N_{PV}$ and $\langle\mu\rangle$ is removed.

The $p_T$ dependence[6] of the pile-up sensitivity is well described by a logarithmic fit, with positive dependence on $N_{PV}$ and negative on $\langle\mu\rangle$. This is however not taken into account in the residual correction, but taken as a systematic uncertainty of the method, contributing to the total JES uncertainty. Apart from uncertainties on $N_{PV}$ and $\langle\mu\rangle$, which enter in the residual correction, a topology uncertainty is included to take bias in $\rho$ from hard jet activity[7] into account.

*Mass Response*

For a given $p_T$, jet mass $m$ increases with the amount of wide-angle radiation. With respect to pile-up, the jet mass is predominantly sensitive to the added clusters from in-time pile-up; in fact it is largely insensitive to out-of-time pile-up. For mass in particular, the area plays a large role. Typical slopes of $p_T$ with $N_{PV}$ are shown in

---

[6]Correcting a dependence of $p_T$ on $p_T$ itself gets less accurate in data, where only the biased $p_T$ is known.

[7]For instance, a $t\bar{t}$ event typically has more high-$p_T$ jets than a $Z \to \mu\mu$ event.

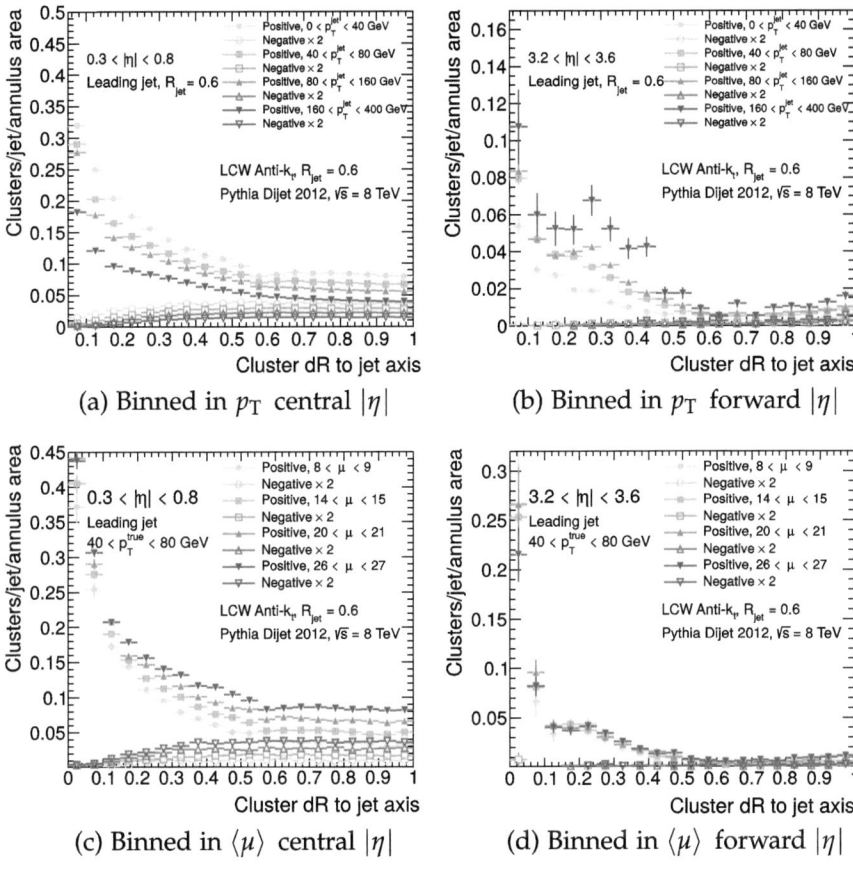

**Fig. 7.11** The distribution of positive and negative clusters in with respect to the leading jet axis, for different $p_T^{lead}$, $|\eta|$ and $\langle\mu\rangle$. **a** Jet axis in the central and **b** forward detector, for different $p_T^{lead}$. **c** Jet axis in the central and **d** forward detector, for different $\langle\mu\rangle$

Fig. 7.13. There is a factor two smaller slope in $m$ than in $p_T$, but it scales with area as given by the two $R$ choices, as before.

The impact of in-time pile-up on the mass distribution is shown in Fig. 7.14. We see in this and the previous figure that the mass of QCD jets can be made independent of pile-up. However, this quantity is foremost interesting as a background in studies of jets stemming from the decay of a massive object, such as a top quark ($t$) or $W$ decay. This was briefly studied using a simulation of fully-hadronic decays of the suggested beyond the SM particle $Z'$,[8] as $Z' \rightarrow t\bar{t}, t \rightarrow Wb, W \rightarrow q\bar{q}$. Depending on the boost, governed by $m_{Z'}$, and the $R$ used for clustering, this can produce a combination of $t$-jets, or a $W$- and a $b$-jet, or two light quark jets and a $b$-jet. Here, $m_{Z'} = 1$ TeV. The identification of hadronically decaying top hinges on recon-

---

[8] A $Z'$ being a heavy $Z$, it decays to the same SM particles.

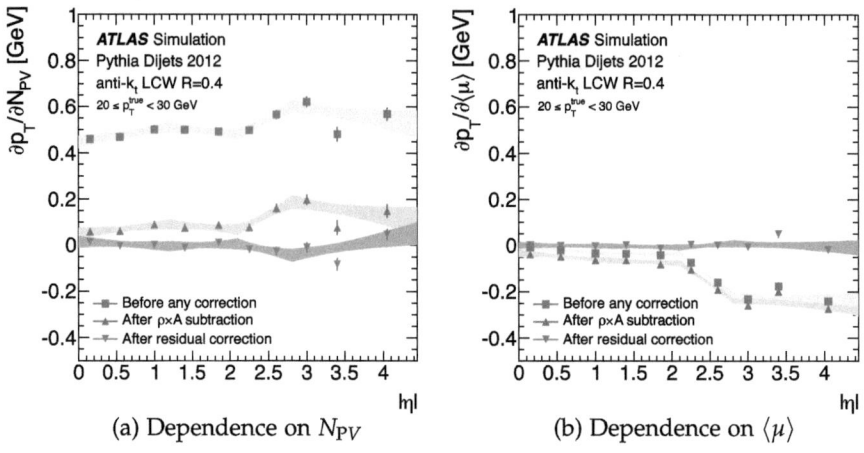

**Fig. 7.12** The dependence of $p_T$ on **a** $N_{PV}$ and **b** $\langle\mu\rangle$, as function of $|\eta|$, before and after $\rho \cdot A$ subtraction, for jets clustered with anti-$k_t$, $R = 0.4$. A residual correction removing the remaining dependence after $\rho \cdot A$ correction is also shown

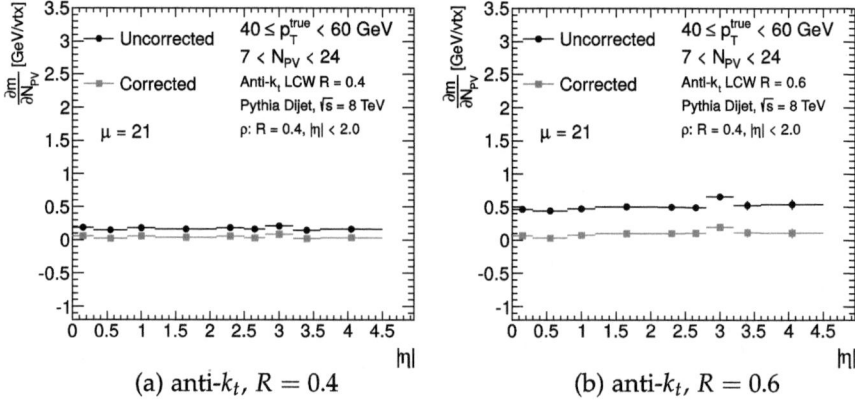

**Fig. 7.13** Dependence on $N_{PV}$ of the jet mass $m$, shown vs $|\eta|$ for anti-$k_t$ jets in the range $40 \leq p_T^{true} < 60\,\text{GeV}$, with **a** $R = 0.4$ and **b** $R = 0.6$

structing the $t$ mass from the resulting jets. Here pile-up smearing will reduce the reconstruction efficiency. Figure 7.15 shows the impact of pile-up on the jet mass distributions, and the result after correction. Two peaks are seen, from $W$ and $t$ jets, respectively.[9] The two peaks are fitted with Gaussians, to loosely identify the 'top' and '$W$' jet masses. The evolution of the location of the mean with $N_{PV}$ is shown in

---

[9]Since pile-up correction is done before the rest of the calibration steps, it was derived on uncalibrated jets. Since the uncalibrated response is less than unity, the jet masses are not expected to completely reach the tabulated particle masses of Table 2.2.

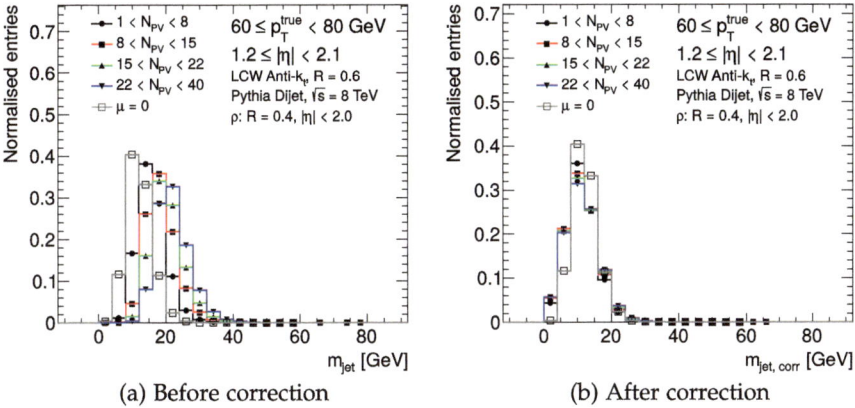

(a) Before correction                                    (b) After correction

**Fig. 7.14** Example distributions of $m_{jet}$ for different $N_{PV}$, before **a** and after **b** correction of the jet four-vector

Fig. 7.15c. It shows that an initial dependence on $N_{PV}$ of the jet mass is removed by the correction.

## 7.3.2 Resolution

### $p_T$ Resolution

The larger sensitivity to fluctuations in pile-up activity in $\rho$ compared to $N_{PV}$ and $\langle\mu\rangle$ manifests itself as a better recovery of the jet $p_T$ resolution compared to the previous method. This is demonstrated in Fig. 7.16. However, quite some dependence on pile-up remains, which is explained by local fluctuations in the pile-up environment in the vicinity of the jet, which are not reflected by $\rho$.

Comparing the two $p_T$ regions shown in Fig. 7.17, we see that the inherent $p_T$ resolution is as expected $p_T$ dependent, but that the pile-up correction reduces the smearing from pile-up equally well in both cases. The detector features in $|\eta|$ are also seen here. It is clear that the pile-up activity fluctuations captured by the centrally measured $\rho$ also hold for the forward region.

### Mass Resolution

Just as for $p_T$, the mass resolution is deteriorated in the presence of pile-up, and partly restored by the correction. This is shown in Fig. 7.18. The improvement is the largest for low-$p_T$ jets. Again an approximate factor two is seen between the pile-up impact on jet mass and $p_T$.

From the shape of the corrected $m$ distributions in Fig. 7.14b, a slight overcorrection can be seen. In fact, there is a fraction of jets, excluded from the above figures,

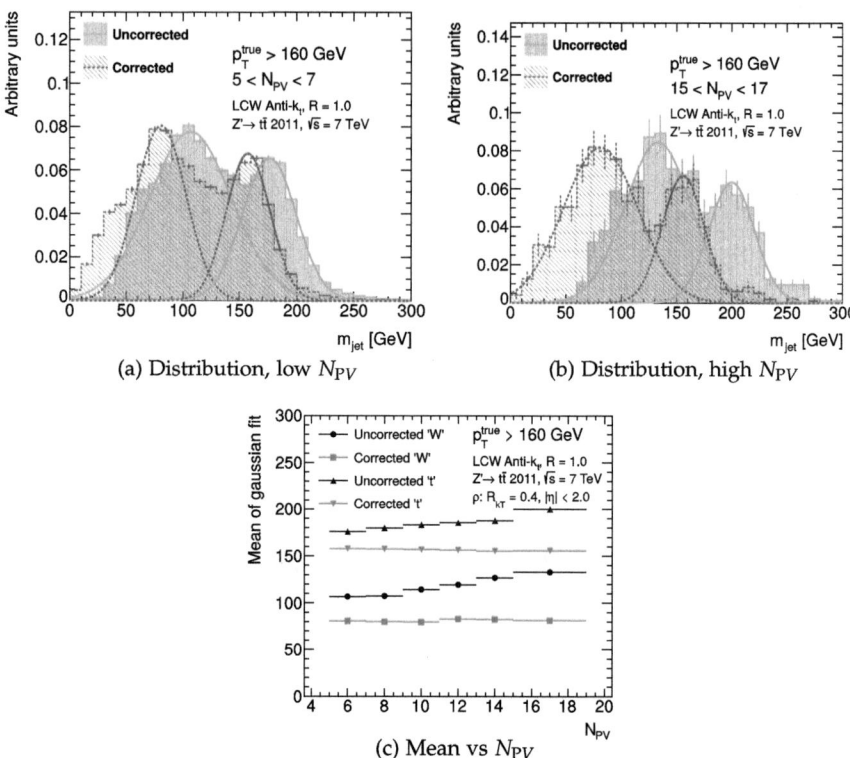

(a) Distribution, low $N_{PV}$          (b) Distribution, high $N_{PV}$

(c) Mean vs $N_{PV}$

**Fig. 7.15** The distributions of $m_{jet}$ in hadronic top decays, for **a** low and **b** high $N_{PV}$, before and after correction. The lines are Gaussian fits. **c** The evolution vs $N_{PV}$ of the mean of the Gaussian fits in the top row

**Fig. 7.16** The dependence on $\langle \mu \rangle$ of the $p_T$ resolution, before any correction, with the previous parameterisation method $f(\langle \mu \rangle, N_{PV})$ and after $\rho \cdot A$ subtraction

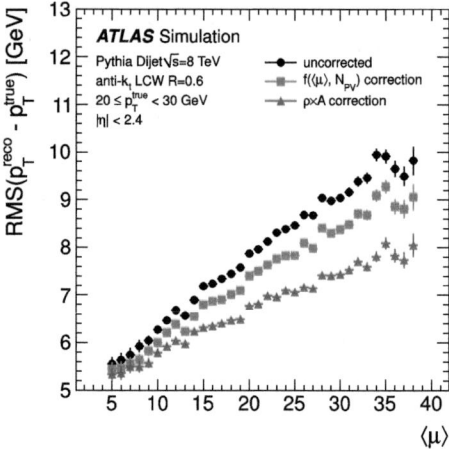

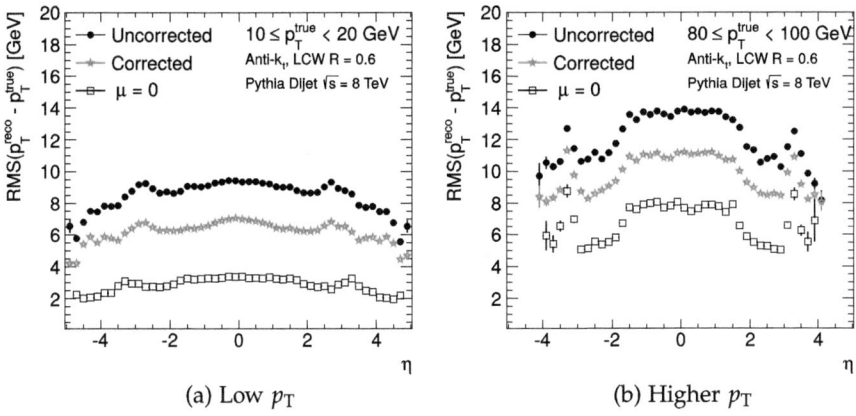

**Fig. 7.17** $p_T$ resolution vs $\eta$, before and after $\rho \cdot A$ subtraction, compared to the resolution in absence of pile-up ($\langle\mu\rangle = 0$), for **a** $10 \leq p_T^{true} < 20\,\text{GeV}$ and **b** $80 \leq p_T^{true} < 100\,\text{GeV}$

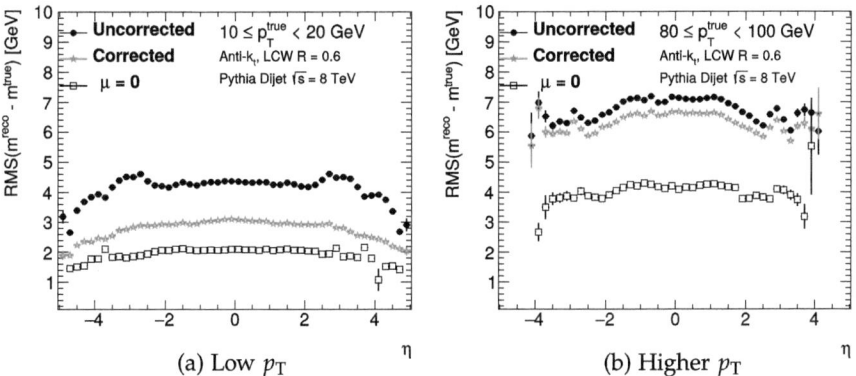

**Fig. 7.18** Mass resolution vs $\eta$, before and after $\rho \cdot A$ subtraction, compared to the resolution in absence of pile-up ($\langle\mu\rangle = 0$), for **a** $10 \leq p_T^{true} < 20\,\text{GeV}$ and **b** $80 \leq p_T^{true} < 100\,\text{GeV}$

getting negative $m^2$ after correction. This is flagged as negative $m$ from a setting in FASTJET:

$$\sqrt{m^2} = \begin{cases} -\sqrt{|m^2|}, & m^2 < 0 \\ \sqrt{m^2}, & m^2 \geq 0 \end{cases} \tag{7.3}$$

where $m^2 < 0 \Rightarrow E^2 < p^2$. Closer studies revealed a few differences between the jets with $m^2 < 0$ and $m^2 \geq 0$. Firstly, jets at the same $p_T^{true}$ and truth jet mass, similar $\rho$, $N_{PV}$ and $|\eta|$ were compared. For jets getting the same reduction in $p_T$, the ones with negative corrected mass showed about a factor two larger mass correction than the positive mass ones. This points to a different topology in the ghosted four-vector area of the jets with negative corrected mass. Secondly, the number of constituent clusters per unit area has a different dependence on $N_{PV}$ as function of $|\eta|$ in the negative-

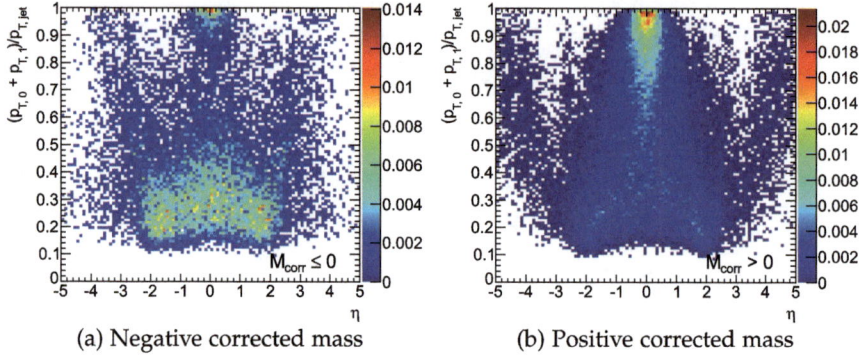

**Fig. 7.19**  The fraction of the jet $p_T$ carried by the two hardest clusters in the jet vs $\eta$, for jets with **a** negative and **b** positive pile-up corrected jet mass

and positive-corrected mass jets: the number decreases with $|\eta|$ as expected for the well-behaved jets, but shows very little $|\eta|$ dependence in the negative-mass ones. Thirdly, the $p_T$ fraction carried by the two highest-$p_T$ clusters in the jet is smaller in the jets with negative corrected masses than in the others. This is illustrated in Fig. 7.19.

Further studies of the jets with negative corrected masses shows that their fraction grows with $|\eta|$ and decreases with $p_T$.[10]

To summarise, it looks as though there are more clusters in these jets, especially at high $|\eta|$, where the impact of pile-up on these jets seems to be different. Something makes the jet area four-vector model the jet four-vector poorly; for instance the high cluster density could relate to differences in the jet area. One conjecture is that these are pile-up jets,[11] possibly from combination of contributions from different pile-up vertices. The problem was not resolved and it was decided to not correct the jet mass using the four-vector area, but to continue using a scale factor relating $p_T$, $E$ and $m$, based on the size of the $p_T$ correction from the $\rho \cdot A$ subtraction.

### 7.3.3  Jet Multiplicity

The final category where pile-up alters observables, is the jet multiplicity. Pile-up can add additional hard-scatter jets originating from a pile-up primary vertex, or diffuse radiation clustered into a pile-up jet, or simply bring jets above some threshold by increasing their $p_T$. Figure 7.20 shows the distribution of the number of jets $N_{jets}$ above 20 GeV in PYTHIA8, in the central region of the detector, in an environment of no pile-up ($\langle \mu \rangle = 0$), compared to jets in a pile-up environment before and after $\rho \cdot A$

---

[10] At high $p_T$, the impact of pile-up *and pile-up correction* gets relatively smaller.

[11] This would explain the $p_T$ regime and different topology.

**Fig. 7.20** The distribution of the number of jets $N_{jets}$ with $p_T$ above 20 GeV within $|\eta| = 2.0$, in the case of no pile-up ($\langle\mu\rangle = 0$), before correction (black) and after $\rho \cdot A$ subtraction (red)

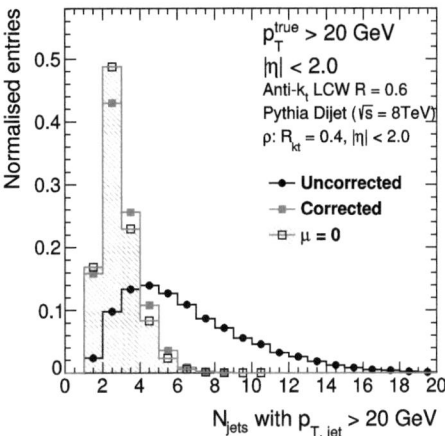

**Fig. 7.21** The dependence on $\langle\mu\rangle$ of the average number of jets $\langle N_{jet}\rangle$ with $p_T$ above 20 GeV within $|\eta| = 2.1$, in $Z \rightarrow \mu\mu$ events in data (filled markers) and MC (open markers), before correction (circles) and after correction (squares)

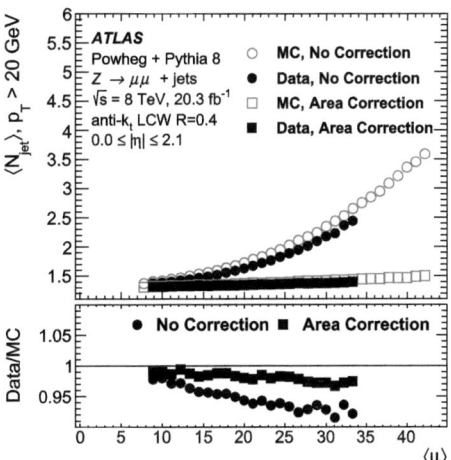

subtraction. The distribution is almost completely restored by the correction, with only a small shift towards higher $N_{jets}$. The residual pile-up dependence is shown in Fig. 7.21, which shows the same quantity as a function of $\langle\mu\rangle$, in $Z \rightarrow \mu\mu$ events in both data and MC. The residual dependence is $\sim 0.02$ jets/vertex, amounting to a reduction by a factor twenty-five. The agreement between data and MC is much improved after correction. This reflects the fact that the additional soft radiation from pile-up is inherently non-perturbative and harder to model than the hard scatter process.

In the central region, the multiplicity of jets depends on pile-up until a lower $p_T$ threshold of 50 GeV. In the forward region, the dependence on pile-up of the jet multiplicity is weaker, and vanishes already at a threshold of 40 GeV.

## 7.4   Potential for Improvements

The pile-up correction method is already supplemented at low $p_T$ by the use of tracks, where several observables exist that identify jets with a large fraction of their tracks associated to pile-up primary vertices. This helps remove the pile-up dependence of jet multiplicity at lower $p_T$ in the central region, where tracking information exists. In the forward region, the need is reduced, but there are strong use cases for jets at low $p_T$ at high rapidity. One example is $H$ boson production through vector boson fusion, producing $H$ bosons with large rapidity boost. At high rapidity, a substantial part of the correction comes from the residual correction parameterised in $N_{PV}$ and $\langle \mu \rangle$, which we know does not capture fluctuations very well. Improving the $\rho \cdot A$ correction in the forward region is thus desirable.

For the forward region, out-of-time pile-up is a substantial challenge. Being able to measure the out-of-time history of an event would be useful here, but storing information on the activity in un-triggered bunch crossings would be impossible considering output rate and event size. However, there is information about the pile-up history in the calorimeter cells themselves—this is what we see as out-of-time pile-up. The information in the negative clusters, that are also formed in topoclustering, has not been exploited for this purpose in the corrections so far. I explored using a $\rho_{neg}$, calculated using negative clusters as input, instead of positive. A combination $\rho = \rho_{pos} + \rho_{neg}$, where $\rho_{neg} < 0$, did reduce the negative $\langle \mu \rangle$ dependence in the forward region. However, the negative clusters aren't calibrated, so using them would require further validation. Considering the interplay of the impact of negative energy and granularity, it might be advantageous to include negative clusters in the region $|\eta| \gtrsim 2.5$ only.

Another possibility would be to implement a correction parameterised in the ratio of negative clusters in the core of the jet to the periphery. This would give jet-by-jet information about the local out-of-time activity, assuming that the negative clusters on the periphery of the jet are a proxy for the amount of overlapping negative energy in the jet core. The $|\eta|$ dependence of this ratio does resemble the impact of out-of-time pile-up on jets (as hinted in Fig. 7.11). The distributions shown in Fig. 7.11 are simple geometrical distributions; one could explore ghost association of negative clusters for a refined matching. Furthermore, splitting the correction into a core-part and an area-dependent part may accurately capture the different regimes of pile-up sensitivity in a jet.

The other obvious candidate for improvement is to resolve the issue with the negative corrected jet masses. In the central region, tracking information can be used to find out if these are hard-scatter jets. For the forward region, exploring the use of a different area definition may be interesting, as the coarse granularity and topoclustering may imply that the catchment area of an anti-$k_t$ jet is smaller than the $\pi R^2$ expectation from this jet algorithm.

Finally, I note that a jet-level pile-up correction can only go part of the way. The optimal pile-up correction would be a constituent-level correction, removing pile-up clusters, and recalibrating the cluster energy for the impact of overlap with out-of-time energy deposits, before jet finding is done. If the impact of pile-up at the calorimeter-level is understood, the need for additional higher-level corrections is reduced—not to mention the increased pile-up robustness in other variables like $E_T^{miss}$ and isolation calculations. While topo-clustering does part of the job, the challenge here is to take the cluster seeding of jets and the negative energy overlap properly into account.

# References

1. ATLAS Collaboration, Pile-up subtraction and suppression for jets in ATLAS. Technical Report ATLAS-CONF-2013-083. Geneva: CERN (2013)
2. ATLAS Collaboration, Performance of pile-up mitigation techniques for jets in pp collisions at $\sqrt{s} = 8$ TeV using the ATLAS detector. Eur. Phys. J. C **76**(11), 581 (2016)
3. M. Cacciari, G.P. Salam, Pileup subtraction using jet areas. Phys. Lett. B **659**, 119–126 (2008)

# Part III
# Dijet Angular Distributions as a Probe of BSM Phenomena

At this point, we have constructed the concept of jets and understood the process of making statements about them based on energy depositions that we read out from our detector. We have seen some of the tools used in jet finding, encountered calorimeter readout and the subtleties arising from overlaid energy stemming from different proton interactions, and appreciated some of the fine details in the many steps of jet calibration.

Now, here we stand with our calibrated jets. It is time to make use of them in a measurement.

# Chapter 8
# Dijet Measurements

## 8.1 Dijet Observables

Once we have an event with a pair of jets, we can form the so-called *dijet* by four-vector addition of the two jet four-vectors. From this new four-vector, we extract the *dijet mass* $m_{jj}$, again using the relation

$$m_{jj}^2 = E_{jj}^2 - p_{jj}^2, \tag{8.1}$$

where the subscript $jj$ means that all quantities are dijet four-vector quantities. From here on, we will assume that the dijet is formed from the two jets with the highest $p_T$ in the event—the *leading* and *subleading* jet.

The invariant mass is particularly interesting precisely because it is invariant—it is a conserved quantity in the case of a particle decay into two jets, for instance. It also probes the energy scale of the collision. These two aspects make the invariant mass an important dijet observable, as an excess of events at a certain $m_{jj}$ can hint at the production of a new heavy particle (possibly seen as a *bump* in a dijet mass spectrum), or generally that we have reached an energy regime where new phenomena become accessible.

The dijet mass spectrum analysis is a sister analysis to the angular distribution analysis, with substantial overlap in phase space and search philosophy. The analysis of the two distributions have often been performed by the same team within ATLAS. While the main focus of this thesis is to analyse the angular distributions, we will come back to the dijet mass spectrum analysis approach and results as we proceed.

© Springer International Publishing AG 2017
L.K. Bryngemark, *Search for New Phenomena in Dijet Angular Distributions at √s = 8 and 13 TeV*, Springer Theses, DOI 10.1007/978-3-319-67346-2_8

### 8.1.1  Dijet Kinematics

In the following[1] we will consider a $2 \rightarrow 2$ process, and assume that we have reconstructed two jets from it. For the kinematics derivation however, we will refer to the partons, and assume that we can transfer the findings to the jets afterwards.

For massive objects, like jets, rapidity as defined in Eq. 3.1 is used, rather than pseudorapidity. In Chap. 3, we introduced $y_B$, the boost of the collision centre-of-mass frame with respect to the detector frame. Denoting the outgoing parton rapidities as $y_1$, $y_2$, this is expressed as

$$y_B = \frac{y_1 + y_2}{2} \tag{8.2}$$

—simply the average rapidity of the two partons. Similarly, we can construct

$$y^* = \frac{y_1 - y_2}{2}, \tag{8.3}$$

where $\pm y^*$ is the rapidity of the two partons in the centre-of-mass frame. The centre-of-mass scattering angle is related to $y^*$ through $\cos \hat{\theta} = \tanh y^*$.

The two outgoing partons must be perfectly balanced in the centre-of-mass frame, where by construction all incoming momenta add up to zero. If initial transverse momenta are neglected, the outgoing partons in a $2 \rightarrow 2$ process will from momentum conservation also balance in $p_T$ and azimuth in the detector frame, with the longitudinal boost, as we saw, giving the transformation between the two frames. From Eqs. 8.2 and 8.3 it is clear that the rapidity difference is independent of choice of longitudinal boost.

### 8.1.2  Angular Distributions, χ

For the same $2 \rightarrow 2$ process with massless particles, we introduce the Mandelstam variables, letting 1 and 2 denote the outgoing parton indices:

$$\hat{s} = (p_1 + p_2)^2$$
$$\hat{t} = -\frac{\hat{s}}{2}\left(1 - \cos\hat{\theta}\right) \tag{8.4}$$
$$\hat{u} = -\frac{\hat{s}}{2}\left(1 + \cos\hat{\theta}\right)$$

---

[1] I am indebted to Nele [1] for the clear explanations of the kinematics and its implications—and for leaving it to me to work out the details…

from which we see that $|\hat{t}| \leq \hat{s}$, $|\hat{u}| \leq \hat{s}$. Being the centre-of-mass energy squared, $\hat{s}$ is related to the proton-proton ($pp$) collision centre-of-mass energy and incoming parton Bjorken $x$ as $\hat{s} = x_i x_j s$.

At leading order, the QCD $2 \rightarrow 2$ scattering process is dominated by $t$-channel exchange, where a gluon is exchanged, resulting predominantly in small scattering angles (where $\hat{t} \rightarrow 0$, $\hat{u} \rightarrow -\hat{s}$). The differential partonic cross section can be expressed in terms of $\hat{t}$ and $\hat{s}$ as

$$\frac{d\hat{\sigma}}{d\hat{t}} \propto \frac{1}{\hat{s}^2} \overline{\sum |M|^2}, \tag{8.5}$$

where in the notation for matrix element squared, $\overline{\sum}$ implies that colour and spin indices are averaged and summed over for the initial and final states, respectively.

For the dominant process at high $p_T$, $qq' \rightarrow qq'$,[2] one finds that

$$\overline{\sum |M|^2} \propto \alpha_s^2 \cdot \frac{\hat{s}^2 + \hat{u}^2}{\hat{t}^2} \sim \alpha_s^2 \cdot \frac{\hat{s}^2}{\hat{t}^2}. \tag{8.6}$$

Using this in Eq. 8.5, we see that

$$\frac{d\hat{\sigma}}{d\hat{t}} \propto \frac{\alpha_s^2}{\hat{t}^2}. \tag{8.7}$$

This equation can be rewritten in two ways to make it more enlightening. Firstly, expressing it in terms of $\cos\hat{\theta}$, and using the expression for $\hat{t}$ from Eq. 8.4, we recover the angular behaviour of Rutherford scattering:

$$\begin{aligned}\frac{d\hat{t}}{d\cos\hat{\theta}} &= \hat{s}/2 \\ \frac{d\hat{\sigma}}{d\cos\hat{\theta}} &\propto \frac{\alpha_s^2}{\hat{t}^2}\frac{\hat{s}}{2} = \frac{2\alpha_s^2}{\hat{s}(1 - \cos\hat{\theta})^2} = \frac{\alpha_s^2}{2\hat{s}\sin^4(\frac{\hat{\theta}}{2})}\end{aligned} \tag{8.8}$$

Introducing the rapidity difference measure $\chi$,

$$\chi = e^{|y_1 - y_2|} = e^{2|y^*|} = \frac{1 + |\cos\hat{\theta}|}{1 - |\cos\hat{\theta}|} \sim \frac{1}{1 - |\cos\hat{\theta}|} \tag{8.9}$$

for $t$-channel processes, using the last equality in Eq. 8.9, the differential cross section becomes

$$\begin{aligned}\chi &\sim -\frac{\hat{s}}{2\hat{t}} \Rightarrow \frac{d\chi}{d\hat{t}} \sim \frac{\hat{s}}{2\hat{t}^2} \\ \frac{d\hat{\sigma}}{d\chi} &\propto \frac{\alpha_s^2/\hat{t}^2}{\hat{s}/\hat{t}^2} = \frac{\alpha_s^2}{\hat{s}}\end{aligned} \tag{8.10}$$

---

[2]cf. the PDFs in Fig. 2.4.

We thus see that for a fixed $\hat{s}$—or experimentally: $m_{jj}$—the differential production cross section $d\hat{\sigma}/d\chi$, is flat as a function of $\chi$ for $t$-channel exchange. This is thus what we expect from lowest order QCD.

A more isotropic event will in turn be independent of $\hat{\theta}$, implying that $\frac{d\hat{\sigma}}{d\cos\hat{\theta}}$ is constant. Using the previous relations, we can derive the following expressions for the differential cross section:

$$
\begin{aligned}
\frac{d\hat{\sigma}}{d\hat{t}} &\propto \frac{1}{\hat{s}} \\
\frac{d\hat{\sigma}}{d\chi} &\propto \frac{\hat{t}^2}{\hat{s}^2} \propto \left(\frac{1}{\chi}\right)^2,
\end{aligned}
\tag{8.11}
$$

meaning that the cross section peaks at low $\chi$.

Many new phenomena are expected to have isotropic distributions: for instance, the distribution of decay products of a new particle produced in the collision. Furthermore, a phenomenon that can be produced at the LHC must couple to partons in some way, meaning, that it can produce partonic final states, which lead to jets. This makes deviations in the dijet angular distributions a good indicator of phenomena beyond the SM. As mentioned in Chap. 1 this possibility has been explored at a range of energies; I have searched for such deviations using the two highest energy data sets to date, with $\sqrt{s} = 8$ and 13 TeV. These two analyses are the focus of the remainder of this thesis.

## 8.2  Tools in the Analysis of Angular Distributions

At LO, we can equate

$$
m_{jj}^2 = \hat{s} = 4p_T^2 \cosh^2(y^*).
\tag{8.12}
$$

Noting that all quantities are positive, and that $\cosh(x) = \cosh(|x|)$, the relationship between $p_T$, $m_{jj}$ and $\chi$ becomes

$$
\begin{aligned}
m_{jj} &= 2p_T \cosh(|y^*|) = p_T(e^{|y^*|} + e^{-|y^*|}) \\
&= p_T\left(\sqrt{\chi} + \frac{1}{\sqrt{\chi}}\right),
\end{aligned}
\tag{8.13}
$$

which provides a very useful intuition: for a given $m_{jj}$, the higher $p_T$ jets are found at low $\chi$.

Going to the hadron level, the differential cross section $\frac{d\sigma}{d\chi}$ is traditionally referred to as the angular distribution. The hadron level cross section is obtained through integration over the momentum fractions and PDFs multiplied by the partonic cross section:

$$\frac{d\sigma}{d\chi} = \int dx_1 \int dx_2 f_1(x_1, Q^2) f_2(x_2, Q^2) \frac{d\hat\sigma}{d\chi} \quad (8.14)$$

To isolate the partonic cross section, $dx_1 dx_2$ can be expressed in terms of $d\tau dy_B$, letting $\tau = x_1 x_2 = \hat{s}/s$, $y_B = \frac{\ln(x_1/x_2)}{2}$. So, $\chi$, $\hat{s}$ and $y_B$ are an equivalent set of parameters for expressing the hadronic cross section. Keeping $\hat{s}$ fixed—experimentally, $m_{jj}$—, the value in a given $\chi$ bin is thus given solely by the partonic cross section, which is independent of the varying $y_B$.[3] However, $y_B$ encodes the convolution with the PDFs, which does not affect the partonic cross section but re-weights the distribution. As discussed in Chap. 2, the PDFs are not calculable directly in QCD, while the partonic cross sections are. Since possible new phenomena modify the partonic cross section, it is desirable to maximise the sensitivity to deviations from QCD predictions, and thus to minimise the influence of the convolutions with the PDFs. Thus a narrow range in $y_B$ is preferable.

### 8.2.1 Comparing the Angular Distributions to Prediction

With the addition of higher orders (and non-perturbative effects), the angular distribution is no longer fully flat in $\chi$, and it needs to be compared to a more elaborate prediction. We have seen that signs of new physics phenomena—angular distributions that are more isotropic than QCD—are expected to appear as deviations at low $\chi$. The analysis of the angular distributions is a comparison of shape in data and SM simulation and relies on this ability to discern such deviations. To minimise the uncertainties in the modelling of the SM distributions, the angular distributions are normalised to have the same integral as data. In visualisation, all angular distributions are normalised to unit area. This highlights the shape and simplifies comparisons across $m_{jj}$ regions.

*Simulation*

The LO generator PYTHIA8 is used to obtain the QCD prediction of the angular distributions.[4] It is a complete generator, simulating the whole process from matrix elements to hadronization. The modelling of non-perturbative effects is subject to tuning of free parameters in QCD, such as $\alpha_s(M_Z)$, the amount of initial and final state radiation, etc. The tuning is made in comparison with experimental data, and thus the non-perturbative effects partly compensate the lack of higher order calculations.

*Corrections*

The LO predictions are brought to NLO accuracy using NLOJET++ [2–4]. This is not a complete generator, but a tool for calculating NLO cross sections for hard processes with up to three-jet final states. It can provide bin-by-bin correction factors

---

[3] The physics does not depend on the choice of reference frame.
[4] More SM simulation details are given in Chap. 9.

obtained from running both LO and NLO calculations. The procedure is described in more detail in Sect. 8.4.

In addition to the QCD corrections, at high energies, EW corrections become important [5]. It can be thought of as the decreasing importance of the mass of the weak bosons: there are contributions from virtual exchange of soft or collinear weak gauge bosons, resulting in Sudakov-type logarithms, with the leading term evolving as $\alpha_W \ln^2(Q^2/M_W^2)$. Tree-level EW corrections of $\mathcal{O}(\alpha_s\alpha, \alpha^2)$ and weak loop corrections at $\mathcal{O}(\alpha_s^2\alpha)$ are provided by the authors of Ref. [5], as bin-by-bin correction factors resulting from cancellation between (positive) tree-level and (negative) 1-loop effects. The tree-level effects are negligible for regimes corresponding to low Bjorken $x$, where gluons dominate. In addition, there is a strong angular trend, since the Sudakov regime requires both $\hat{s}$ and $\hat{t}$ to be large. As we have already seen, $\hat{t} \to 0$ for small scattering angles, i.e., high $\chi$. It is shown in Ref. [5] that the tree-level contributions are also larger at low $\chi$ due to interference effects.[5]

### 8.2.2  Statistical Analysis

The statistical analysis is based on doing a shape comparison between SM prediction and data, obtained through normalising the SM prediction to have the same integral as data. The distribution used is the number of events $N$ vs $\chi$,[6] for a given $m_{jj}$. From here, a test statistic can be formed, used to test the compatibility of the observed data with both the null hypothesis that it follows the SM prediction, and the alternate hypothesis that it follows the distribution predicted by a combination of SM and signal processes.

We can in general express the observed number of events in terms of a signal strength $\mu$, with $0 \leq \mu \leq 1$, as

$$N = \mu S + B, \tag{8.15}$$

where $S$ is the nominal number of signal events predicted by the signal model, and $B$ is the number of background events, here the prediction by SM.

*Statistics tool*

The number of selected events coming from a potential new physics process of cross section $\sigma$ is $N_{new} = L \times \sigma \times A \times \epsilon$, where $L$ is the integrated luminosity and $A \times \epsilon$ is the product of the acceptance and efficiency of the event selection criteria. We note that in the nomenclature of Eq. 8.15, $N_{new} = 1 \cdot S$, i.e., with $\mu = 1$. Upper limits on $\sigma \times A \times \epsilon$ relate the maximum $N_{new}$ (or equivalently, $\mu$) still compatible with the data at 95% confidence level.

---

[5]The argument is that they are dominated by $qq$ initiated processes. It is stated that the interference terms don't receive contributions from the squares of the $t$-, $u$- and $s$-channel diagrams, while these dominate the (forward) LO QCD contributions.

[6]Technical point: since the analysis framework expects equidistant binning, $\ln(\chi)$ is used.

From the input distributions for a given $m_{jj}$ region, a Poisson likelihood model for the description of the event yield is constructed: if the expected number of signal and SM events in bin $i$ are $s_i$ and $b_i$, respectively, the likelihood for the distribution in $n$ bins become

$$P(\text{data}|\text{SM} + \text{signal}) = \prod_{i=1}^{n} \frac{(s_i + b_i)^{n_i} (e^{-(s_i+b_i)})}{n_i!}, \tag{8.16}$$

where $n_i$ is the number of events in bin $i$, and again $s_i$ is related to the nominal signal prediction for that bin as $s_i = \mu S_i$. The corresponding SM-only likelihood is obtained with $s_i = 0$ ($\mu = 0$). The likelihood ratio $q_\mu$ can be formed as

$$q_\mu = \frac{P(\text{data}|\text{SM} + \text{signal})}{P(\text{data}|\text{SM})}. \tag{8.17}$$

The Modified Frequentist method ($CL_s$ method) [6] is then used to extract the upper limit on $q_\mu$. Here we will use the signal strength throughout, but this could also be some other parameter of the model, for instance the cross section itself. Systematic uncertainties on predicted signal and background yields, as well as the (anti)correlation of their effects on the distribution shape, are taken into account by adding a nuisance parameter $\hat{\theta}$ for each source of uncertainty. In a profile likelihood, $\hat{\theta}$ is the value maximising the likelihood from a Gaussian likelihood $G(\text{data}|\theta, \sigma_\theta)$, and is obtained in a simultaneous fit of the prediction to data, along with all the nuisance parameters. The systematic uncertainties are described in Sect. 10.5.

For each tested value of $\mu$, the variable

$$\begin{aligned}
\text{CL}_s &= \frac{p_\mu}{1 - p_b} = \frac{P(q_\mu \leq q_\mu^{\text{obs}}|\text{SM} + \text{signal})}{P(q_\mu < q_\mu^{\text{obs}}|\text{SM})} \\
&= \frac{\int_{q_\mu^{\text{obs}}}^{\infty} f(q_\mu|\mu, \hat{\theta}_\mu) dq_\mu}{\int_{-\infty}^{q_\mu^{\text{obs}}} f(q_\mu|0, \hat{\theta}_0) dq_\mu}
\end{aligned} \tag{8.18}$$

is computed, as the ratio of the integrals from the observed value of the test statistic, $q_\mu^{\text{obs}}$, to infinity, of the probability density functions $f(q_\mu)$ when the true value of the parameter of interest is either the tested signal $\mu$ ($p_\mu$) or zero (SM-only hypothesis, $p_b$). The 95% CL limit on $\mu$ is then given by the solution to the equation $\text{CL}_s = 0.05$. The compatibility of the data with the null hypothesis is reported as $\text{CL}_b = 1 - p_b$, which from unitarity of probability corresponds to the integral in the denominator taken from $q_\mu^{\text{obs}}$ to positive infinity. Analytic asymptotic formulae describing the test statistic distributions $f$ are used for the results shown in this thesis, as described in Ref. [7]. It has been verified that the asymptotic formulae give the same solution as a sampling of the test statistic distribution from toy experiments.

## 8.3  Binning Considerations

### 8.3.1  $\chi$ Binning

Following the calorimeter granularity, a rapidity binning in multiples of the typical Tile calorimeter cell width of 0.1 minimises smearing and modulations from edge effects. Defining the location of the $\chi$ bin edge $i$ as $\exp(0.3 \times i)$ was demonstrated to be optimal [1], using 11 bins with the last bin edge visually extended to $\chi = 30$, with the actual event selection cut at $y^* = 1.7$ ($\chi \approx 29.94$). Especially considering the wide bins used for $\chi$, migrations due to the angular resolution of jets are negligible [8, Appendix X].

The definition of the $\chi$ binning makes the bin width grow exponentially. For visualisation, logarithmic horizontal axes are used for $\chi$, highlighting the more interesting low $\chi$ region. One should however keep in mind that the distribution is dominated by the high $\chi$ region, for instance in the normalisation procedure, making signal bias negligible.

### 8.3.2  $m_{jj}$ Binning

Apart from stabilising the angular distributions, as shown above, introducing a binning in $m_{jj}$ also gives a sensitivity to the scale of possibly emerging new phenomena at high $m_{jj}$. The $d\sigma/d\chi$ distribution is binned coarsely in $m_{jj}$ using bins of $\mathcal{O}(100\,\text{GeV}-1\,\text{TeV})$. The lowest possible $m_{jj}$ is dictated by the $p_T$ thresholds used in the event selection, which in turn are typically derived from the trigger efficiency, which has a turn-on curve. Only triggers at 99.5% or higher efficiency are used, which avoids efficiency corrections as well as matching schemes. A lowest order indication of the resulting $m_{jj}$ threshold corresponding to a $p_T$ threshold is given by Eq. 8.13, inserting the maximum $\chi$ used, but in practice it is often slightly higher.

The width of the $m_{jj}$ intervals used are subject to optimisation of the balance between statistical and systematic uncertainty, but also sensitivity to benchmark signals, and in the case of a first-data search,[7] flexibility! The optimisation of the cut values is detailed in Chap. 10.

## 8.4  NLO QCD Corrections: $K$-factors

PYTHIA calculates hard scattering processes to LO only, but since it is a complete event generator, some of the missing higher order processes are partially made up

---

[7]For a fast search, the final data set is not known when the analysis design is laid out, and there needs to be room for later changes.

for in the parton showering. It can be thought of as a "partial" NLO correction $K_{part}$ which is already factored into the PYTHIA8 simulation. As NLO perturbative calculations are more precise, it is advantageous to correct the PYTHIA prediction to NLO by applying bin-wise $K$-factors.[8] These corrections are derived as a ratio between the NLO and LO cross section for the hard process, calculated using NLOJET++. In the procedure to apply the NLO corrections, care must be taken to handle the non-perturbative contributions in PYTHIA correctly. For this a dedicated PYTHIA8 sample with only hard process and parton showering turned on,[9] $LO^{PYTHIA}_{show}$, is used. One can then define an NLO corrected PYTHIA8 prediction as

$$\text{PYTHIA}_{corr} = \frac{(NLO/LO)^{\text{NLOJET++}}}{(LO_{show}/LO)^{\text{PYTHIA}}} \cdot \text{PYTHIA}_{reco}, \tag{8.19}$$

where we have identified $K_{part} = (LO_{show}/LO)^{\text{PYTHIA}}$ and divided the PYTHIA prediction by it. Assuming $LO^{\text{NLOJET++}} = LO^{\text{PYTHIA}}$, we define a $K$-factor from

$$\text{PYTHIA}_{corr} = \frac{NLO^{\text{NLOJET++}}}{LO^{\text{PYTHIA}}_{show}} \cdot \text{PYTHIA}_{reco} \equiv K \cdot \text{PYTHIA}_{reco} \tag{8.20}$$

The assumption that the two LO predictions are equal relies on recognising that these are the pure QCD matrix element predictions. The only thing which can introduce a difference between them is using a different choice of PDF and $\alpha_s$ in the calculations, so the calculation requires using the same PDF in NLOJET++ as in PYTHIA8. Note that the PDF sets have to match the order of the calculation.

The calculations for the $\sqrt{s} = 8$ TeV analysis are explicitly shown here. The CT10 PDF set is used, at the proper order in QCD, but there is a difference remaining in the choice for $\alpha_s$. This is set differently in the calculations in the two generators, as they use the evolution of $\alpha_s$ at different orders: $\alpha_s^{\text{NLOJET++}}(M_Z) = 0.118$ and $\alpha_s^{\text{PYTHIA8}}(M_Z) = 0.135$.

To illustrate the impact of this difference, all components needed to obtain the correction factor in Eq. 8.19 are drawn in Fig. 8.1(a), showing the differential cross section $d\sigma/d\chi$ in the dijet mass bin $2000 \leq m_{jj} < 2600$ GeV, as predicted from the LO matrix element calculations of both generators, as well as the NLO prediction from NLOJET++ and the PYTHIA8 prediction where non-perturbative effects are turned off. In Fig. 8.1(b), the PYTHIA LO prediction has been rescaled by the square of the ratio of the two values of $\alpha_s$, taken at $p_T^{avg} = Q = 350$ GeV, which following Eq. 8.13 is the minimum $p_T$ required for a dijet mass in this range. The running of $\alpha_s$ follows the LO evolution corresponding to the first term in Eq. 2.8, with the number of light flavours $n_f$ taken to be 5. The value of $\Lambda_{QCD}$ is a matter of tuning, and

---

[8]For angular distributions, one example that would introduce deviations from the $2 \to 2$ kinematics could be an additional final state jet. This is an NLO hard-scatter process, but could also be achieved at LO by hard final-state radiation off of one jet in a $2 \to 2$ scattering. Both cases would modify the angular separation of the hardest two jets in the system from the back-to-back case.

[9]The PYTHIA8 settings to stop after parton showering are given in Appendix A.

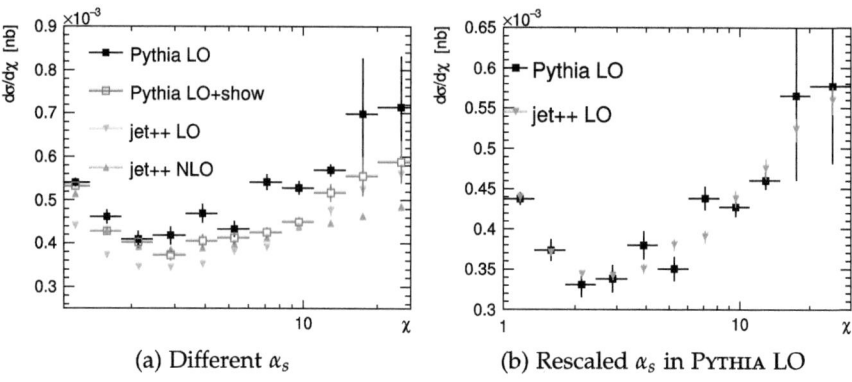

**Fig. 8.1** Predicted differential cross section $d\sigma/d\chi$ at $\sqrt{s} = 8$ TeV in the dijet mass bin $2000 \leq m_{jj} < 2600$ GeV, using **(a)** different $\alpha_s$ and **(b)** rescaling the PYTHIA prediction to the $\alpha_s$ value used in NLOJET++

known to be $\Lambda_{QCD} = 0.2262$ GeV in NLOJET++. It is left floating in the PYTHIA case to fit the set value of $\alpha_s^{PYTHIA8}(M_Z)$ used in the $\alpha_s$ evolution. The values obtained by this evolution are in agreement with those given in [9] for different values of $Q$.

It is clear from Fig. 8.1 that the difference in the two LO predictions is introduced solely from the difference in $\alpha_s$, as it vanishes once the rescaling of PYTHIA LO is done. It is also clear, that this difference merely introduces a shift in the normalisation between the two predictions. As the final analysis uses distributions normalised to the data integral, the difference introduced by the different default $\alpha_s$ values vanishes, and one can safely neglect the LO predictions in the $K$-factor expression. For each $m_{jj}$ region, a $\chi$ dependent $K$-factor is thus calculated as

$$K(\chi) = \frac{NLO^{\text{NLOJET++}}(\chi)}{LO^{\text{PYTHIA}}_{show}(\chi)} \tag{8.21}$$

It is also evident in Fig. 8.1(a) that the parton showering process in PYTHIA8 affects the shape of the $\chi$ distribution in a similar way as the addition of higher orders, but doesn't give the full answer. Comparing the shapes, we can expect $K$-factors with only a small dependence on $\chi$. These are then applied to the default LO PYTHIA8 sample with hard process, parton showering, multiple interactions, and non-perturbative effects turned on. The result is a NLO partonic prediction that has been corrected with non-perturbative effects. These $K$-factors are applied before normalisation of the $\chi$ distributions.

Finally, the prediction from PYTHIA8 with $K$-factors applied is compared to a POWHEG [10–12] prediction of the angular distributions of QCD. POWHEG predicts the hardest emissions to NLO accuracy and is then interfaced to PYTHIA8 for showering. Albeit at NLO, it is not used as the default QCD prediction, since, firstly

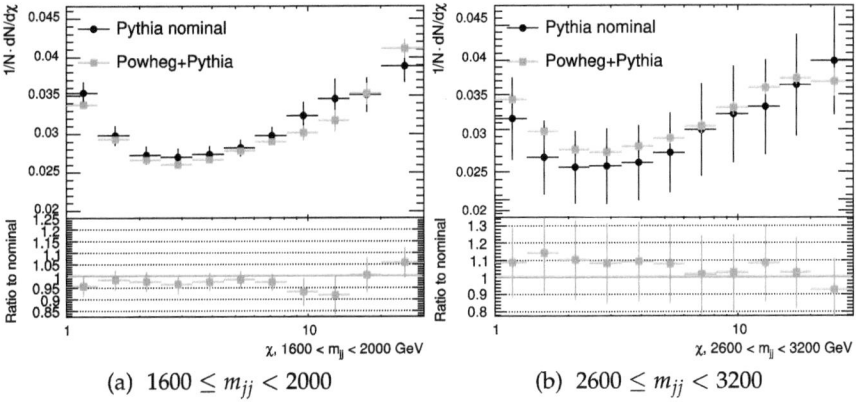

(a) $1600 \leq m_{jj} < 2000$        (b) $2600 \leq m_{jj} < 3200$

**Fig. 8.2** Comparison of the $\sqrt{s} = 8\,\text{TeV}$ differential cross section $d\sigma/d\chi$ (normalised to unit area) predicted by POWHEG to PYTHIA with $K$-factors, in the dijet mass bin (**a**) $1600 \leq m_{jj} < 2000\,\text{GeV}$ and (**b**) $2600 \leq m_{jj} < 3200\,\text{GeV}$

the CI signal prediction is obtained with PYTHIA[10] (see Chap. 9 for more details on the signal samples) and secondly, the available $\sqrt{s} = 8\,\text{TeV}$ POWHEG sample was smaller which introduces large statistical fluctuations, particularly at high $m_{jj}$. Still, the POWHEG sample serves as a reference of what the NLO prediction would be. Figure 8.2 shows the comparison of the POWHEG and PYTHIA8 $\times$ $K$-factors predictions at reconstructed level, for two example bins in $m_{jj}$. The ratio of the two is consistent with 1 within errors for all bins. The normalisation is somewhat driven by the last bin in $\chi$, which is the widest, sometimes introducing large shifts between the predictions even though the general shape agrees.

## 8.5 Dijet Mass Distribution

For the dijet mass distribution, the same kinematics hold, and the sensitivity to isotropic events is enhanced by selecting only low-$\chi$ events. Given the relation in Eq. 8.13, this implies a lower reach in $m_{jj}$ for a given $p_T$. The angular distributions thus probe the highest $m_{jj}$ events produced, while the dijet mass distribution displays the $m_{jj}$ evolution of the low $\chi$ region, which is expected to be the signal-enriched region in the angular distributions.

The underlying assumption in the design of a prediction for the SM dijet mass spectrum is that it is featureless: in the absence of new scales, it is simply smoothly— and rapidly—falling. This is exploited in a data-driven background estimate, fitting a smooth parameterisation to the data. This gives a search with only one systematic

[10]In particular, the signal has interference with QCD, which means having the same QCD prediction in both background and signal modelling is easier.

uncertainty: the choice of fit function. The parameterisation and the method used to choose the fit will be described along with the distributions in Chap. 11.

The slight differences in the analysis strategy of the two distributions does entail differences in their sensitivity. Firstly, while the small systematic uncertainties in the SM prediction of the dijet mass distribution yields a large sensitivity to deviations from the prediction, the prediction itself may become sensitive to signal. For instance, the typical $m_{jj}$ evolution of CI is an onset of a modified cross section that gets more pronounced with higher $m_{jj}$. This affects the tail of the $m_{jj}$ distribution. A smooth fit of sufficiently many parameters would easily accommodate this change in shape. This makes the dijet mass spectrum particularly suited to search for narrow resonances, locally enhancing the dijet cross section, but less so for non-resonant phenomena. Secondly, the dijet mass measurement is a rate measurement, and it uses the lack of an increased rate expected from signal to set limits on the cross section of a hypothesised new process. This is different from analysing the angular distributions which is above all a shape measurement.[11]

# References

1. N. Boelaert, Dijet angular distributions in proton-proton collisions at $\sqrt{s} = 7$ TeV and $\sqrt{s} = 14$ TeV, Presented on 21 Sep 2010. Ph.D. thesis, Lund University (2012)
2. Z. Nagy, Three-jet cross sections in hadron-hadron collisions at next-to-leading-order. Phys. Rev. Lett. **88**, 122003 (2002)
3. Z. Nagy, Next-to-leading order calculation of three-jet observables in hadron-hadron collision. Phys. Rev. D **68**, 094002 (2003)
4. S. Catani, M.H. Seymour, A general algorithm for calculating jet cross-sections in NLO QCD. Nucl. Phys. B **485**, 291–419 (1997)
5. S. Dittmaier, A. Huss, C. Speckner, Weak radiative corrections to dijet production at hadron colliders. JHEP **1211**, 095 (2012)
6. A.L. Read, Presentation of search results: the CL(s) technique. J. Phys. G **28**, 2693–2704 (2002)
7. G. Cowan et al., Asymptotic formulae for likelihood-based tests of new physics. Eur. Phys. J. **C71**, 1554 (2011)
8. A. Ashkenazi et al., ATLAS search in 2012 data for new phenomena in dijet mass distributions using pp collisions at $\sqrt{s} = 8$ TeV (2013)
9. M. Lichtnecker, Determination of as via the Differential 2- Jet-Rate with ATLAS at LHC. Ph.D. thesis. Ludwig-Maximilians- Universität München (2011)
10. S. Frixione, P. Nason, C. Oleari, Matching NLO QCD computations with parton shower simulations: the POWHEG method. JHEP **0711**, 070 (2007)
11. S. Alioli et al., A general framework for implementing NLO calculations in shower Monte Carlo programs: the POWHEG BOX. JHEP **1006**, 043 (2010)
12. S. Alioli et al., Jet pair production in POWHEG. JHEP **1104**, 081 (2011)

---

[11]Of course, the cross section does enter in the fact that a signal needs to be discernible even in the presence of SM processes.

# Chapter 9
# Signal Model Sample Generation

This chapter describes the modelling of the SM prediction and the signals introduced in Sect. 2.8 in more detail.

## 9.1 QCD

The baseline SM prediction in the angular distributions is the NLO QCD and EW corrected PYTHIA8 prediction. Being a leading order generator, it implements the calculations outlined in Chap. 8, which provides a suitable starting point for dijet production simulation.

In the $\sqrt{s} = 8$ TeV analysis, the AU2 [1] underlying event tune and leading-order CT10 [2] PDFs are used. At $\sqrt{s} = 13$ TeV, ATLAS had moved to the A14 underlying event tune [3] and leading-order NNPDF2.3 [4, 5] PDFs.

It is known for both the $\sqrt{s} = 8$ and 13 TeV MC samples that the PYTHIA cross section prediction at high $m_{jj}$ is 20–30 % too large compared to data. This is attributed to a shift in the $m_{jj}$ distribution (the spectrum is harder in MC than in data). However, the prediction from PYTHIA6.421 [6] with underlying event tune AUET2B [1] had shown better cross section and shape agreement with ATLAS data at $\sqrt{s} = 7$ TeV.

To investigate the cause of the shift, a set of PYTHIA8 truth-level samples with variations of for instance $\Lambda_{QCD}$ and the amount of initial-state radiation [7–9] was compared to the prediction from PYTHIA6.423 [6] with underlying event tune AUET2B. It was verified that the baseline PYTHIA8 prediction at fully reconstructed level deviated only very slightly from the distributions at truth level, meaning that the truth-level tunes could be compared to reconstructed MC without loss of validity. The disagreement between cross section in data and MC was shown to be present already at parton level; the tune variations could not explain the difference seen between PYTHIA6 and PYTHIA8.

© Springer International Publishing AG 2017

L.K. Bryngemark, *Search for New Phenomena in Dijet Angular Distributions at $\sqrt{s} = 8$ and 13 TeV*, Springer Theses, DOI 10.1007/978-3-319-67346-2_9

## 9.2   Contact Interactions

CI is used as benchmark signal in both the $\sqrt{s} = 8$ and 13 TeV searches. In the modelling of CI, only left-chiral colour singlet coupling is considered (corresponding to the subscript $L$ in Eq. 2.10), meaning $\eta_{LL} = \pm 1$, $\eta_{RR} = \eta_{RL} = 0$. This is a simplification whose predictions already cover most of the range obtained by considering also right-handed states. The signal is generated with PYTHIA8, along with QCD, modelling interference with the SM process $q\bar{q} \rightarrow q\bar{q}$. Thus the same PDF set and PYTHIA8 tune is used for CI as for QCD modelling. The branching ratio to quarks is 100%. Example $\chi$ distributions of these generated samples at $\Lambda = 7$ and 10 TeV are shown in Fig. 9.1. The signal strength increases with $m_{jj}$.

### 9.2.1   $\Lambda$ Scaling

Apart from the $m_{jj}$ range at which it is probed, the total signal cross section depends on interference mode and the scale $\Lambda$, and a modelling of this evolution requires a scan in these 2 dimensions. However, the CI amplitude (from here on called "CI$^2$ term") and interference term scale as $1/\Lambda^4$ and $1/\Lambda^2$ respectively, meaning that the cross section can be obtained at an arbitrary $\Lambda$ by rescaling. A large sample for each interference mode was generated at $\Lambda = 7$ and 10 TeV, with the 10 TeV sample used for validation of the extrapolation procedure, which is as follows. By adding the resulting histograms from the two interference modes at a given $\Lambda$, the interference terms cancel, and only $2 \times (\sigma_{QCD} + \sigma_{CI^2})$ terms remain. Since the QCD term is known from the SM prediction MC, the CI$^2$ term can be isolated. Similarly, by subtraction of the two samples, the interference term is isolated. The obtained $\chi$ distributions of signal and interference terms for $\Lambda = 7$ TeV are shown for $3.2 \leq m_{jj} < 8.0$ TeV in Fig. 9.2.

Using the dependence on $\Lambda$ for the two terms respectively, full rescaling of the simulated samples can be achieved for any value of $\Lambda$. The validity of this rescaling has been tested by comparing the $\Lambda = 10$ TeV distributions obtained from the rescaled $\Lambda = 7$ TeV samples, to those from the generated $\Lambda = 10$ TeV sample. The two versions agree to within 4% (with the largest deviations in the high-$\chi$ region, where the signal contribution is negligible and thus the statistical uncertainty is large), as shown in Fig. 9.3.

### 9.2.2   Signal K-factors

Also the signal is brought to NLO precision using $K$-factors. The LO and NLO cross section for each $m_{jj}$ region and $\chi$ bin is derived using the CIJET package (v 1.0) [10] for the destructive and constructive interference term and for all the

**Fig. 9.1** $\chi$ distributions of signal and QCD generated at $\sqrt{s} = 8$ TeV for different dijet invariant mass ranges

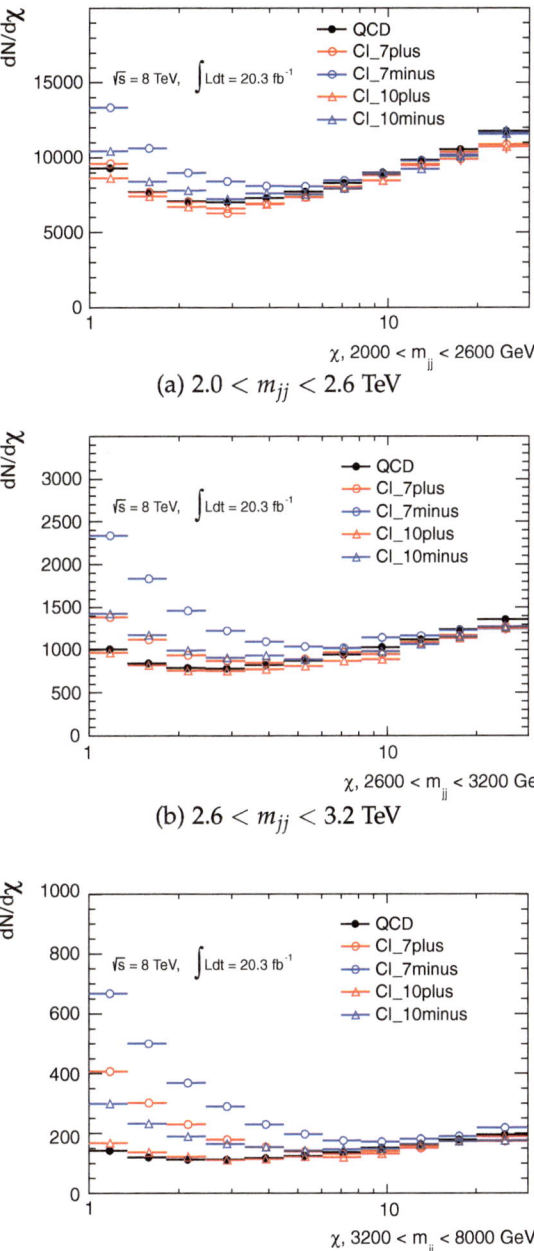

(a) $2.0 < m_{jj} < 2.6$ TeV

(b) $2.6 < m_{jj} < 3.2$ TeV

(c) $3.2 < m_{jj} < 8.0$ TeV

**Fig. 9.2** χ distributions of
signal CI and interference
terms for Λ = 7 TeV and
3.2 < $m_{jj}$ < 8.0 TeV

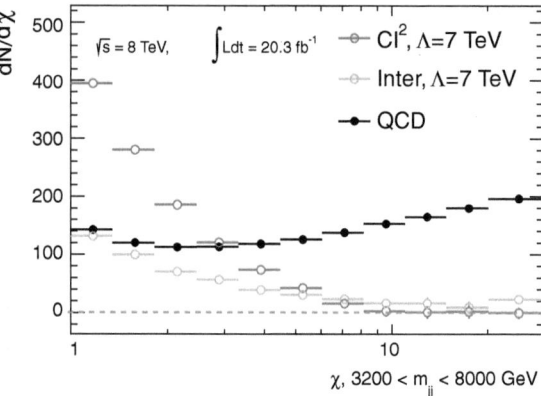

**Fig. 9.2** χ distributions of
signal CI and interference
terms for Λ = 7 TeV and
3.2 < $m_{jj}$ < 8.0 TeV

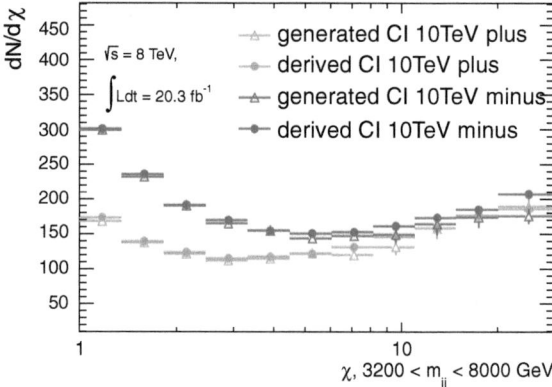

**Fig. 9.3** Comparison of
signal distribution of Λ = 10
TeV between generated and
derived signal prediction

extrapolated values of Λ. As before,[1] by adding and subtracting the LO (NLO)
contributions from the positive and negative interference term, one can obtain the
LO (NLO) contribution of the $CI^2$ term and the interference term separately for each
bin. Dividing the NLO CI (interference) term by the corresponding LO contribution
will give the $K$-factors for each bin, as shown in Fig. 9.4. This figure shows that the CI
$K$-factors are overall less than unity, and that the interference term $K$-factor has the
strongest χ dependence. After applying these $K$-factors, we obtain the final signal
contributions at various Λ scales, shown in Fig. 9.5. There are no EW corrections for
signal corresponding to those available for the QCD simulation.

The spacing between the curves in Fig. 9.4 evolves very smoothly with Λ. Indeed,
when drawn for each χ bin as function of Λ, as shown for one region in $m_{jj}$ at
$\sqrt{s} = 13$ TeV in Fig. 9.6, a similar shape vs Λ is seen for all χ bins. A tentative
parameterisation $K = p_0 + \frac{p_1}{\Lambda}$ (dashed line in the figures) shows good promise
for finding an evolution with Λ that can be used for both inter- and extrapolation,

---

[1] The only difference is that there is no QCD term to be subtracted.

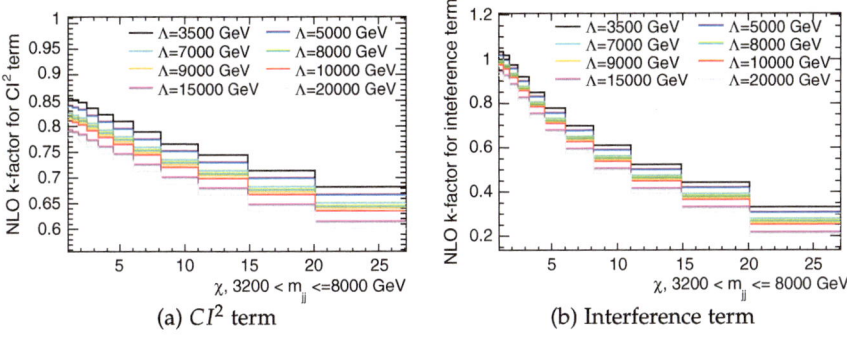

**Fig. 9.4** NLO $K$-factor vs $\chi$ for CI signal in $3.2 < m_{jj} < 8.0$ TeV, calculated for $\sqrt{s} = 8$ TeV

**Fig. 9.5** $\chi$ distributions of signal with NLO $K$-factor applied for various $\Lambda$ scales

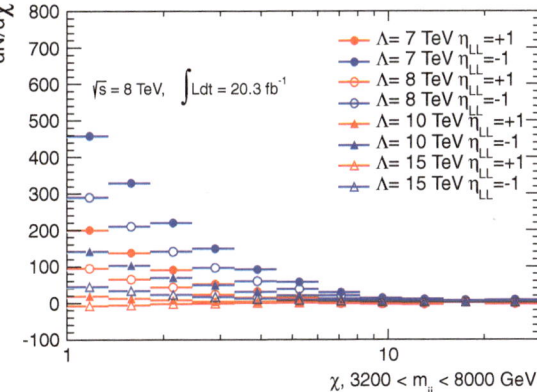

since running CIJET is time consuming. Such an approach would complement the flexibility in the signal template extrapolation described above.[2]

## 9.2.3 Normalisation

The CI prediction is generated along with QCD, and the QCD prediction from PYTHIA is subtracted. As we have discussed, the PYTHIA prediction of the cross section is too high compared to data. The cause of this is unknown, even after extensive tune studies for the 8 TeV MC prediction. However, since it is present already at parton level, and since QCD and CI are generated together, it is assumed that to lowest order, the over-prediction of the cross section in QCD is present also in the pure CI

---

[2]One can note, however, that the evolution with $\Lambda$ is slow, so the gain in precision from short-range extrapolation is small.

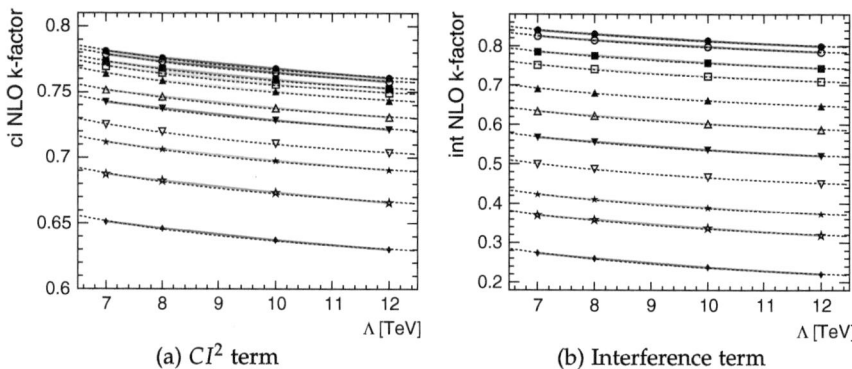

**Fig. 9.6** NLO $K$-factor vs $\Lambda$ for CI signal in $3.4 < m_{jj} < 3.7$ TeV, calculated for $\sqrt{s} = 13$ TeV. Each curve corresponds to a bin in $\chi$, with the lowest one at the top. The dashed line is a fit

prediction. Thus the CI signal nominally corresponding to the integrated luminosity in the data is rescaled by the same factor $N_{\text{data}}/N_{\text{QCD MC}}$ as the QCD prediction.

## 9.3  Quantum Black Holes

QBH is used as an example resonant phenomenon, with a broad peak in $m_{jj}$ and still with a distinct angular distribution compared to the SM. The QBH signal is characterised by a wide resonance of large cross section, and can be discovered or excluded in a range beyond the $m_{jj}$ reach expected from the SM at a given integrated luminosity. Its cross section is large enough to make a visible signal also in a wide $m_{jj}$ bin like the last $m_{jj}$ region used for the angular distributions.

Two different realisations of the $n = 6$, ADD model black holes, produced at the threshold mass $M_{th}$ equal to the fundamental scale of gravity $M_D$, are used as benchmarks in the analysis of the angular distributions recorded at $\sqrt{s} = 13$ TeV. They are implemented by two different generators: BlackMax [11] and QBH [12].[3]

BlackMax implements a sophisticated decay model taking into account the possibilities of black holes with rotation from a non-zero impact parameter, or recoiling into the bulk; split-fermion branes (that quarks and leptons exist on different 3D branes); non-thermal or thermal decay, etc. Decays in the QBH generator, in turn, are dictated by the local gauge symmetries of the SM. Despite the modelling differences, the final $m_{jj}$ distributions from the two generators largely have the same shape while the cross section differs. The branching ratio to states producing dijets is 96%.

---

[3]Since the generator and model name are the same, we will use the convention to denote the model with QBH and the generator with QBH.

## 9.4  Excited Quarks

The narrow resonance benchmark model samples of $q^*$ signal templates are generated for different masses $m_{q^*}$ using PYTHIA8, with the A14 tune [3] and NNPDF2.3 PDF set [5]. The $qg \rightarrow q^*$ production model [13, 14] is used, with the assumption of spin 1/2 and quark-like SM coupling constants. The compositeness scale ($\Lambda$) is set equal to $m_{q^*}$. Only decays to a gluon and light ($u$ and $d$) quarks are simulated, corresponding to a branching fraction of 85%.

## References

1. ATLAS Collaboration, Summary of ATLAS Pythia 8 tunes. Technical Report ATL-PHYS-PUB-2012-003. Geneva: CERN (2012)
2. H.-L. Lai et al., New parton distributions for collider physics. Phys. Rev. D **82**, 074024 (2010)
3. ATLAS Collaboration, ATLAS Run 1 Pythia8 tunes. Technical Report ATL-PHYS-PUB-2014-021. Geneva: CERN (2014)
4. C.S. Deans, Progress in the NNPDF global analysis, in *Proceedings, 48th Rencontres de Moriond on QCD and High Energy Interactions* (2013), pp. 353–356
5. S. Carrazza, S. Forte, J. Rojo, Parton distributions and event generators, in *Proceedings, 43rd International Symposium on Multiparticle Dynamics (ISMD 13)* (2013), pp. 89–096
6. T. Sjöstrand, S. Mrenna, P. Skands, PYTHIA 6.4 physics and manual. JHEP **05**, 026 (2006)
7. P. Skands, S. Carrazza, J. Rojo, Tuning PYTHIA 8.1: the Monash 2013 Tune. Eur. Phys. J. C **74.8**, 3024 (2014)
8. R. Corke, T. Sjöstrand, Interleaved Parton showers and tuning prospects. JHEP **03**, 032 (2011)
9. Y. Bai, A. Katz, B. Tweedie, Pulling out all the stops: searching for RPV SUSY with stop-Jets. JHEP **01**, 040 (2014)
10. J. Gao, CIJET: a program for computation of jet cross sections induced by quark contact interactions at hadron colliders. Comput. Phys. Commun. **184**, 2362–2366 (2013)
11. D.-C. Dai et al., BlackMax: a black-hole event generator with rotation, recoil, split branes, and brane tension. Phys. Rev. D **77**, 076007 (2008)
12. D.M. Gingrich, Quantum black holes with charge, color and spin at the LHC. J. Phys. G: Nucl. Particle Phys. **37**(10), 105008 (2010)
13. U. Baur, I. Hinchliffe, D. Zeppenfeld, Excited quark production at hadron colliders. Int. J. Mod. Phys. A **2**, 1285 (1987)
14. U. Baur, M. Spira, P.M. Zerwas, Excited quark and lepton production at hadron colliders. Phys. Rev. D **42**, 815–825 (1990)

# Chapter 10
# Analysis of Angular Distributions at $\sqrt{s} = 8$ and 13 TeV

This chapter describes the event and data quality selections needed for analysis, as well as the corrections to MC and its systematic uncertainties. Since they are very similar, the $\sqrt{s} = 8$ and 13 TeV analyses [1–3] are described in parallel. This also highlights the occasions where different choices have been made, and a motivation is given.

One might wonder about the motivation for doing the same search twice, shortly after each other. The integrated luminosity used in the two searches differs by an order of magnitude, with the more recent $\sqrt{s} = 13\,TeV$ data set being the smaller one. The answer is, that even with a small data set, we rapidly break new ground if the centre-of-mass energy increases. The reason is the increase in parton luminosity, as illustrated in Fig. 10.1. This figure shows the ratio of the calculated parton luminosities at the two centre-of-mass energies explored in this work, as function of the probed mass. Already at masses around 2 TeV, the penalty from the one order of magnitude smaller integrated luminosity in the $\sqrt{s} = 13$ TeV data set is overcome.

## 10.1 Event Selection

The analysis idea is to measure the dijet angular distributions in events with two or more jets above a certain $p_T$ threshold given by experimental considerations, in a rapidity range that allows a long enough lever arm in $\chi$ for the shape comparison between data and prediction. The different experimental conditions between the $\sqrt{s} = 8$ and 13 TeV analyses warrant a few differences that will be discussed separately below. The overall selection common choices are listed here:

- The highest $\sum_{track} p_T^2$ vertex has at least two tracks associated with it (primary vertex definition)
- Trigger: passes OR of the relevant Level 1 and HLT triggers

© Springer International Publishing AG 2017
L.K. Bryngemark, *Search for New Phenomena in Dijet Angular Distributions at √s = 8 and 13 TeV*, Springer Theses, DOI 10.1007/978-3-319-67346-2_10

**Fig. 10.1** The ratio of
parton luminosities at
$\sqrt{s} = 13$ to 8 TeV [4]

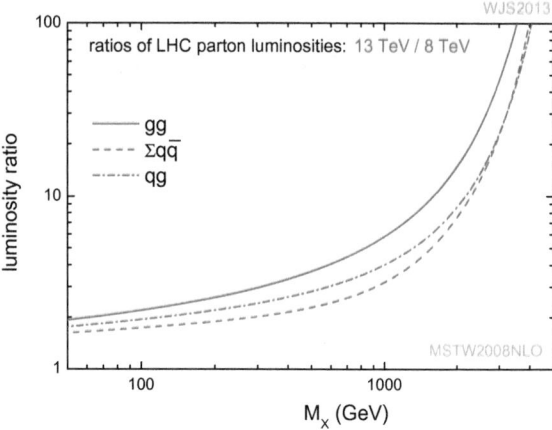

- at least two clean jets with $p_T > 50$ GeV
- Leading jet $p_T > p_T^{thr}$ specific to the two searches
- $|y^*| < 1.7^1$
- $|y_B| < 1.1$
- $m_{jj} > m_{jj}^{thr}$ specific to the two searches

The $p_T$ cut at 50 GeV is governed by pile-up considerations: the number of jets above this cut after pile-up correction is largely independent of pile-up. If the subsequent leading jet $p_T$ cut is high, this lower cut has very marginal impact. It does however remove pathologically unbalanced topologies where the subleading jet $p_T$ is very low, for instance where the leading jet originates from a noise burst or non-collision background. Similarly, the jet cleaning (see Sect. 6.3.2) removes very few additional events since the event selection itself, requiring two jets, removes most events with fake high-$p_T$ jets.

The choice of jet distance parameter differs between the $\sqrt{s} = 8$ and 13 TeV analyses: $R = 0.6$ in the analysis of $\sqrt{s} = 8$ TeV data, while in the analysis of $\sqrt{s} = 13$ TeV data, $R = 0.4$ was used. The larger distance parameter was intended to improve the mass resolution by catching more of the final-state radiation. However, in a study comparing the two distance parameters, no improvement in $m_{q^*}$ resolution was seen. In the interest of harmonisation with other analyses in ATLAS, and based on the knowledge that anti-$k_t$ jets with $R = 0.4$ would be the first jet collection where corrections and uncertainties would become available, it was decided to use the smaller distance parameter in the analysis of $\sqrt{s} = 13$ TeV data. In addition, since the trigger uses anti-$k_t$ jets with $R = 0.4$, a smaller $R$ gives fully efficient triggers at lower reconstructed $p_T$.

Figure 10.2 shows some key variables after the above selection in the 13 TeV data set and PYTHIA8 MC, where all distributions have been reweighted using the ($\chi$, $m_{jj}$)

---

[1]For $m_{jj}$ distributions, this cut is at 0.6. In addition there is no cut on $y_B$.

**Fig. 10.2** A few of the observables, shown in data and MC normalised to the data integral after the selection listed above: the **a** leading jet $p_T$, **b** subleading jet $p_T$, and **c** $m_{jj}$. The JES uncertainty, described later, is shown in shaded blue

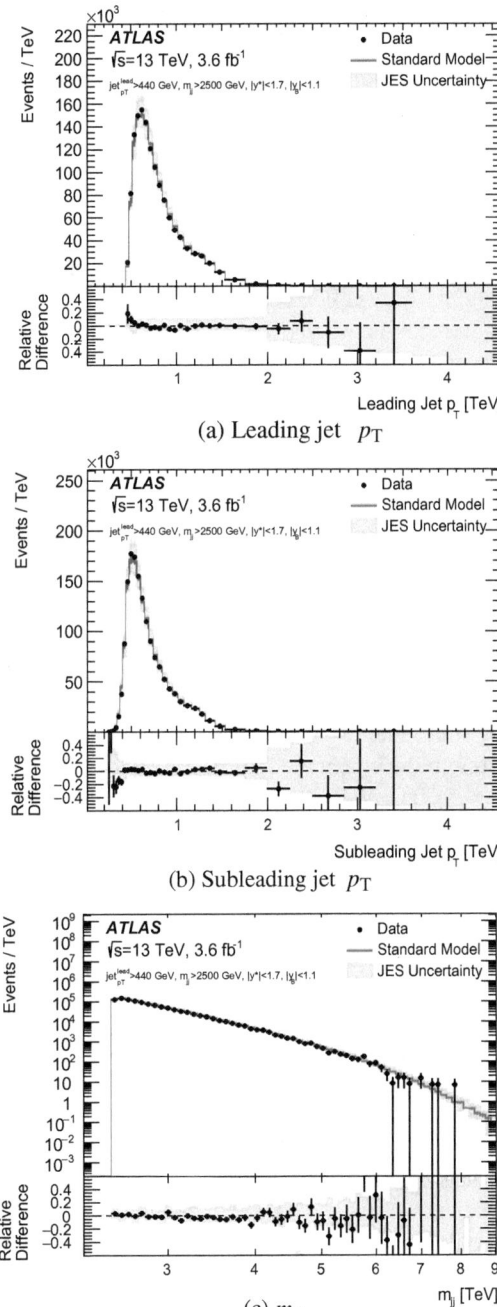

(a) Leading jet $p_T$

(b) Subleading jet $p_T$

(c) $m_{jj}$

dependent NLO QCD $K$- and EW $\kappa$-factors, and MC has been renormalised by a scale factor to match the number of events in data. This scale factor is 0.7, meaning that the MC cross section is $\sim$30% too high in this region of phase space.

### 10.1.1 $\sqrt{s} = 13$ TeV

The analysis of data at $\sqrt{s} = 13$ TeV was done twice: first with the subset of 2015 data that was collected, processed and available for analysis in time for the LHCP conference in August 2015 [2], and then with the full 2015 data set, published in December [3]. The only change in method between the two lies in the statistical analysis, where the full data set uses a more advanced method combining several $m_{jj}$ regions. The earlier search follows the procedure used in the analysis of the $\sqrt{s} = 8$ TeV data, which will be described later.

**Trigger**

The focus of the data analysis at $\sqrt{s} = 13$ TeV was to produce a fast result giving the first look at the new energy regime opening up. Thus, for simplicity, the lowest un-prescaled single jet trigger was used, meaning that all events have an equal statistical weight. The conjecture was that the same hardware and high-level trigger would remain un-prescaled throughout 2015. The trigger $p_T$ threshold at 360 GeV corresponds to full efficiency for anti-$k_t$ $R = 0.4$ jets at 409 GeV, with minimal $|y^*|$ dependence. The corresponding $m_{jj}$ efficiency curve is highly $|y^*|$ dependent, through the kinematic bias described by Eq. 8.12. For $|y^*| < 1.7$, the minimum $m_{jj}$ is $\sim$2000 GeV. However, considering the possibility that the trigger prescales could change, a safety margin was introduced, giving an $m_{jj}$ cut at 2500 GeV.[2]

### 10.1.2 $\sqrt{s} = 8$ TeV

The analysis of the 8 TeV data was done on a mature and final data set, with all conditions already well understood. It was done after the dijet mass resonance search had already been performed on a data set partially overlapping with the phase space of the angular distribution search (with a cut on $|y^*| < 0.6$, allowing for a lower $m_{jj}$ cut, at 250 GeV). This data set was more fully explored, using a combination of triggers reaching all the way down to the pile-up limitation on jet $p_T$. With full efficiency at $m_{jj} = 500$ GeV, the lower mass cut was set at 600 GeV.

**Trigger**

A lower $p_T$ cut of 50 GeV was used for both leading and subleading jet.[3] This is far down in the prescaled regime of the jet $p_T$ spectrum, meaning that the events will

---

[2]For mass distributions, the corresponding cut is at $m_{jj} = 1100$ GeV.

[3]In practice this means that the actual leading jet $p_T$ cut was slightly higher.

have different weights, based on the prescale. For the combination of two triggers with different prescales, let's imagine we have trigger $A$ with a prescale factor of 100, and trigger $B$ with 10, where $A$ is fully efficient at a lower $p_T$ than $B$. Even though trigger $A$ is fully efficient before trigger $B$, in the presence of prescales, it will not record the same events. If there are 100 events above the threshold of full efficiency of trigger $B$, it will have fired 10 times, while $A$ has fired once, and this event may or may not be present also in the set that was triggered by $B$. In the event weight calculation, the probability that a trigger *didn't* fire is used to avoid double counting [5]. The weight is given by:

$$w = \frac{1}{1 - \prod_i \left(1 - \frac{1}{\langle p_i \rangle}\right)}$$

where the index $i$ denotes one of the available fully efficient triggers and $\langle p_i \rangle$ is the average prescale of the single jet trigger $i$.[4] The average prescale reflects the fact that prescales can change in the course of data taking, as the instantaneous luminosity delivered by LHC evolves over time.

In Run1, the offline computing capacity for storing and processing data promptly limited the EF output rate to approximately 400 Hz. The data stream resulting from all EF triggers, and recorded for prompt processing, is called the *normal stream*. In addition, at approximately 200 Hz, ATLAS stored events for later reconstruction. This is called the *delayed stream*, partly derived from a different set of triggers. In the analysis of $\sqrt{s} = 8$ TeV data, it was advantageous to use the delayed stream only, since the lowest un-prescaled trigger had a lower $p_T$ threshold in the delayed stream than in the normal stream. This stream was not active in the beginning of 2012 data taking, reducing the available integrated luminosity from 20.2 to 17.3 fb$^{-1}$.

## 10.2  Corrections

### 10.2.1  Theoretical Corrections

The $m_{jj}$ binning used in the derivation of the following corrections will be discussed shortly (see Sect. 10.4). Figure 10.3 shows the NLO $K$-factors described in Sect. 8.4, derived for the $\sqrt{s} = 13$ TeV MC prediction as function of $\chi$ and for all dijet mass regions, in exclusive binning. The generator settings are given in Appendix A. The $K$-factors become increasingly important with $m_{jj}$. Although a large number of events were generated, there are statistical fluctuations in the $K$-factors which have been reduced by a smoothing procedure,[5] which brings outliers closer to the overall trend using medians of sequences. The edge points and statistical uncertainties

---

[4] If there were only one trigger, the event weight would be $\langle p_i \rangle$.

[5] The ROOT method TH1::Smooth() is used, which employs the 353QH smoothing algorithm, using the repeated median of intervals (3, 5, 3 bins wide) and also includes the smoothed residuals [6].

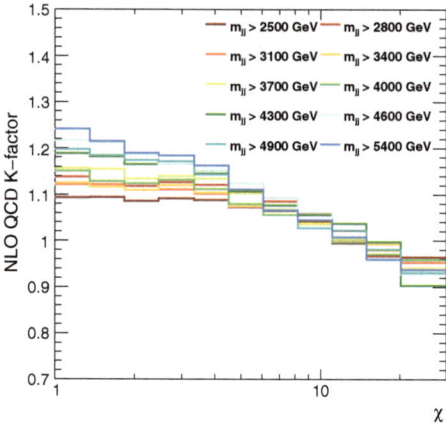

**Fig. 10.3** Shape $K$-factors for the $\chi$ distribution for all dijet mass regions in the analysis of $\sqrt{s} = 13$ data, obtained from distributions normalised to the same integral

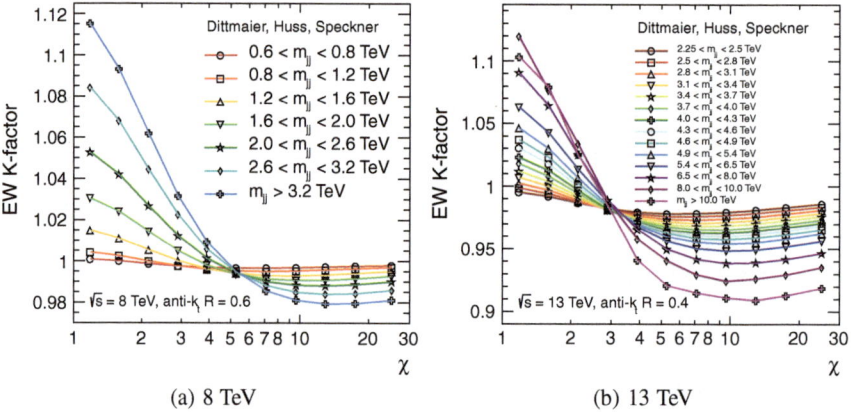

(a) 8 TeV                   (b) 13 TeV

**Fig. 10.4** EW correction $\kappa$-factors for the $\chi$ distribution for all dijet mass bins for **a** $\sqrt{s} = 8$ TeV and **b** $\sqrt{s} = 13$ TeV. The lines are a linear interpolation, there to guide the eye only

are unaffected. The shift of points introduced by the smoothing is well within the statistical uncertainty for all points. The $K$-factor statistical uncertainty is taken as a flat systematic uncertainty, corresponding to the largest uncertainty assessed from NLOJET++.

Figure 10.4 shows the EW correction $\kappa$-factors, provided by the authors of [7]. They have a large effect on the shape of the distributions, bringing the low-$\chi$ region up. The $m_{jj}$ dependence is again clear.

For both centre-of-mass energies, the $\kappa$-factors tend to pivot around the same $\chi$ point in all mass regions, around $\chi = 5$ and 3, respectively. That this pivoting point moves in $\chi$ relates to probing a different region of the PDF. The theorists providing the corrections call the pivoting an "accidental" cancellation. However, I note that

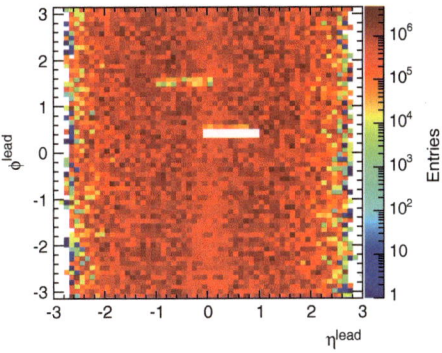

**Fig. 10.5** Impact of removing the masked modules on the leading jet $\eta - \phi$ distribution. The two modules masked for the majority of the data taking are clearly visible

indeed the $\sqrt{s}$ ratio $8/13 \approx 3/5$, which is the relation we would expect if the location in $\chi$ can be mapped from $x_i x_j$.

The statistical uncertainty on the $\kappa$-factors has been made negligible by using large samples. Since they are derived as a relative correction, using the same simulation settings as the nominal MC prediction, no additional systematic uncertainties are taken into account for the EW corrections.

## 10.2.2  Experimental Corrections: Removal of Masked Modules

During the course of 2012 data taking, some modules in the Tile calorimeter were, transiently or permanently, non-responsive. The corresponding cells were masked, meaning that the deposited energy was not read out, leading to a temporary or permanent hole in the solid angle coverage of the calorimeter. An algorithm was implemented to estimate the unrecorded energy based on interpolation between the neighbouring calorimeter cells. Unfortunately, this led to an overestimate of the energy deposited, which would bias the dijet mass measurement towards higher masses. Instead, all events were discarded where one of the two leading jets, or any other jet with $p_T > 0.3 \cdot p_T^{sublead}$,[6] fell into a masked module. This procedure corresponds to a decrease in event rate, since the entire event was discarded. Figure 10.5 demonstrates the impact on the $\eta - \phi$ distribution of the leading jet. However, the distribution in $|\eta|$ of masked modules was non-uniform, with a larger number in the central detector,[7] affecting the overall angular distribution shape with a larger deficit at low $\chi$. The effect thus had to be reproduced in MC.

---

[6]Like for cleaning, this is based on the finding that below this fraction, the maximum energy smearing in a masked module would not introduce changes in the ordering of jets.

[7]cf. the $|\eta|$ coverage of the Tile calorimeter, Chapter 5.

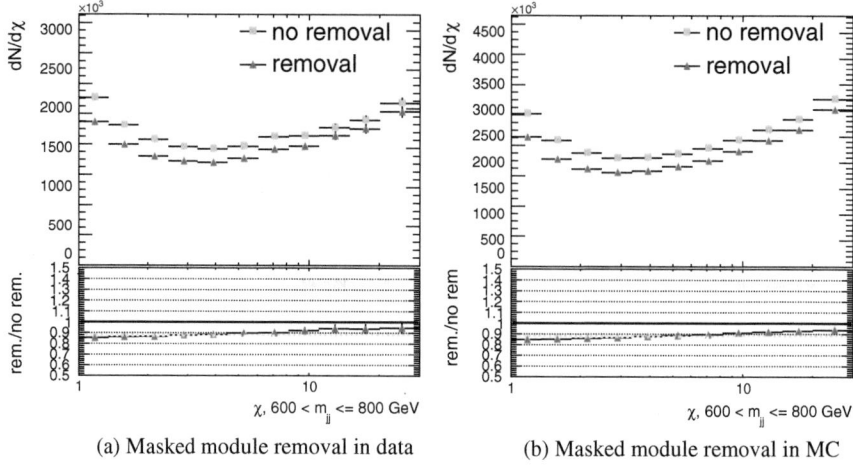

**Fig. 10.6** Overlay of the angular distribution for $600 < m_{jj} \leq 800\,\text{GeV}$ with and without removing the masked modules in **a** data and **b** MC

The dijet event simulation was done before the 2012 data taking was concluded, and only the two permanently masked modules known at the time of MC sample production were included in the detector simulation. This also implies that the MC isn't fully unbiased by this effect, making a MC-based correction of data implausible. In data, more modules were masked with time, introducing a time dependence, translated to a fraction of the integrated luminosity being discarded due to each one of them. The impact of removing the masked modules is mimicked in MC by, for each simulated event, drawing a run number from a sampling distribution which reflects the integrated luminosity collected in each run in the data, and rejecting events where the simulated leading jets fall into those regions which have non-operational Tile modules in the corresponding data taking run. Using the integrated luminosity distribution as described ensures that a representative number of events is discarded in MC for each non-operational module in data.

Figure 10.6 shows the impact on an example angular distribution, for $600 < m_{jj} \leq 800\,\text{GeV}$, of the removal of the masked modules. An average rate reduction of $10 - 15\%$ is seen, affecting the shape[8] of the angular distribution as more events are removed at low $\chi$. The effect is very similar in data and MC, indicating that either the impact of the overcorrection relating to the masked modules is small, or that the two masked modules included in the detector simulation dominate.

---

[8]Although most of the shift is removed in the normalisation, and only the shape change remains, the rate decrease reduces the statistical power of the data.

## 10.3   Statistical Analysis

### 10.3.1   Input

As input for the statistical analysis, the distributions obtained in data, SM and signal predictions at the integrated data luminosity, and the corresponding uncertainties, are used. The pure CI signal distributions, including interference, are obtained for a range of $\Lambda$ with the described extrapolation method, while for QBH, a range of $M_{th}$ is probed. The signal histograms thus obtained corresponds to the expected number of signal events at signal strength $\mu = 1$. In addition to shifting the $m_{jj}$ scale at which the phenomenon appears, the expected signal yield decreases with $m_{q^*}$, $M_{th}$ and $\Lambda$, as exemplified for CI in Fig. 11.1.

### 10.3.2   Procedure

In practice, the statistical analysis proceeds using a HistFactory [8] and RooStats [9] framework, which uses the input histograms to find the best estimate[9] of both the parameter of interest, $\mu$, and the nuisance parameters, given data. Expected limits, which are a measure of the sensitivity of the search to a phenomenon, are derived from testing the compatibility of the signal + SM hypothesis with the SM-only histogram, finding the 95% CL limit of $\mu$ allowed by the statistical and systematical uncertainties. The limit on $m_{q^*}$, $M_{th}$ or $\Lambda$ corresponds to where we can exclude the nominal signal prediction, with $\mu = 1$, at 95% CL. In the event that there is no signal template exactly corresponding to this point, an interpolation between the signal strengths of the two simulated templates straddling $\mu = 1$ is done. For observed limits, the procedure is the same, but the compatibility of signal + SM with the data is tested.

## 10.4   Binning Optimisation

As described in Sect. 8.3.1, the binning in $\chi$ is governed by the detector granularity. The binning in $m_{jj}$ instead has to be optimised with the search sensitivity in mind. Here it is important to remember that we don't know what awaits us; in a broad search for new phenomena like the one described here, one can't afford to tailor the event selection and $m_{jj}$ binning too much towards a specific signal model.

The optimisation of the $m_{jj}$ binning followed slightly different logic in the $\sqrt{s} = 8$ and 13 TeV searches, and will be outlined separately.

---

[9]The best estimate: the value maximising the likelihood.

## 10.4.1 $\sqrt{s} = 8$ TeV

For the $\sqrt{s} = 8$ TeV search, initially two angular variables were explored: $\chi$ and $F_\chi$,

$$F_\chi(m_{jj}) = \frac{N_{|y^*|<0.6}(m_{jj})}{N_{|y^*|<1.7}(m_{jj})}, \tag{10.1}$$

where the $m_{jj}$ binning of $F_\chi$ followed the dijet mass spectrum binning, in turn optimised with respect to the detector $m_{jj}$ resolution. This distribution is thus finely binned in $m_{jj}$ but coarsely in $y^*$ (or, correspondingly, $\chi$), which yields an enhanced sensitivity to resonant phenomena, but still follows the same logic of shape comparison and sensitivity to isotropic phenomena: an increase at low $\chi$ would give an increase in $F_\chi$ too. Thus the angular distributions in $\chi$ were made coarsely binned in $m_{jj}$, using the fine binning in $\chi$ (as described in Sect. 8.3.1), for complementarity. Ultimately, the analysis of $F_\chi$ was not pursued, from a lack of a statistical modelling compatible with the statistics tools used. We will come back to this later.

The analysis of $\sqrt{s} = 8$ data was done in two steps: first a partial data set, consisting of one quarter of the collected events and restricting the $m_{jj}$ range to below 2 TeV, was used for analysis optimisation. This $m_{jj}$ range defined the *control region*, while $m_{jj} \geq 2$ TeV was the "blinded" *signal region*. The signal region was split into three subranges in $m_{jj}$, with boundaries at 2.6 and 3.2 TeV. The $3.2 \leq m_{jj} < 8.0$ TeV range had shown an optimal sensitivity to CI signal, assessed through the expected limits with varied lower boundaries. Given its non-resonant behaviour, the CI signal increases with $m_{jj}$ once it turns on, and optimal sensitivity in the highest $m_{jj}$ region is expected. The remaining two regions were used for a separate assessment of the compatibility of data and SM prediction in each region (testing the null hypothesis), but not for obtaining limits on $\Lambda$. Once all analysis choices were settled, based on good agreement between data and MC in the control region, all data were included, and a statistical analysis was performed in the signal region to establish their compatibility with the SM prediction and the benchmark models.

## 10.4.2 $\sqrt{s} = 13$ TeV

The analysis of $\sqrt{s} = 13$ TeV data was designed and all choices frozen before 2015 data taking started. This approach avoided the need for a control region and a blinded signal region, enabling immediate data analysis for the sake of speed. The optimal $m_{jj}$ binning, however, depends on the size of the data set used, which was not known beforehand. Simulation studies showed that expected reach in $m_{jj}$ was up to $\sim$7 TeV with 10 fb$^{-1}$. Furthermore, with an expected integrated luminosity in the range $1 - 10$ fb$^{-1}$, and with systematic uncertainties of the same size as in the $\sqrt{s} = 8$ TeV analysis, the size of systematic and statistical uncertainty would stay comparable in the last two bins using a bin width of $\sim$300 GeV, except for at the highest $m_{jj}$ where wider ranges were needed. Thus a preliminary narrow binning was chosen, with 300 GeV wide bins starting at the lower $m_{jj}$ threshold of 2.5 TeV, but wider towards high $m_{jj}$. These were used in all derivations of systematic uncertainties and corrections.[10]

Sensitivity studies showed that for a range of integrated luminosities up to 1 fb$^{-1}$, the best sensitivity to CI was obtained in the region from 3.4 TeV up. For the first result at $\sqrt{s} = 13$ TeV, only 80 pb$^{-1}$ of data was used, making this the final bin used. For the analysis of the full 2015 data set, a statistical analysis using a simultaneous fit of 600 GeV wide bins starting at 3.4 TeV had been shown to give clear sensitivity improvements.

When correlating uncertainties across regions of different statistical power, such as low-$m_{jj}$ regions where the statistical uncertainty is low and the large-uncertainty regions at high $m_{jj}$, one needs to verify that nuisance parameters don't get over-constrained by the region of high statistical power. That said, the regions with high statistical power can also provide information about the "true" value of a nuisance parameter, since the central value is optimised—*profiled*—in the fit. For instance, it was seen in Fig. 6.4a that the region of highest jet $p_T$ suffers from large statistical uncertainty, which inflates the JES uncertainty for these jets, starting after $p_T \sim 2$ TeV. A first measurement of this region of phase space can be used to constrain this nuisance parameter by comparing to the actual data, instead of relying on the JES uncertainty previously derived from a smaller population of jets at these energies. It was verified that correlating across regions of different statistical power did not introduce any overconstraints on nuisance parameters, but did move some of their central values slightly, in accordance with the direction of better agreement between data and MC.

The combined fit gives increased sensitivity to resonant phenomena, since it takes the evolution of signal with $m_{jj}$ into account. Sensitivity to narrow resonances benefits from narrower $m_{jj}$ ranges, since the signal is less washed out by the SM background. The sensitivity to resonant signals is exemplified using a $q^*$ signal.[11] Examples of results from a combined fit across the $m_{jj}$ regions in the range 2.5–5.4 TeV is shown in Fig. 10.7.

This type of figure encodes a lot of information. First of all, the left vertical axis, representing the signal strength, is the factor the signal needs to be multiplied by for the experiment to be able to exclude it at 95% confidence level. The horizontal axis denotes the $m_{q^*}$ of the signal hypothesis probed in each $CL_s$ fit. The resulting expected and observed upper limit on signal strength from each fit, along with $1\sigma$ and $2\sigma$ confidence level bands, are combined into this figure. The expected and observed lower limit on $m_{q^*}$ is found where the corresponding lines cross the line at signal strength $\mu = 1$, where the expected limit is interpreted as the sensitivity of the experiment. Here, the sensitivity depends on the $m_{jj}$ range used in the combined fit. Since this range stops at 5.4 TeV in this example, and the signal is narrow in $m_{jj}$, there is not much signal from higher $m_{q^*}$ that makes it into this range, and the experiment

---

[10]The exact binning can be seen in for instance Fig. 10.4, showing the EW corrections.

[11]Details in Appendix A.

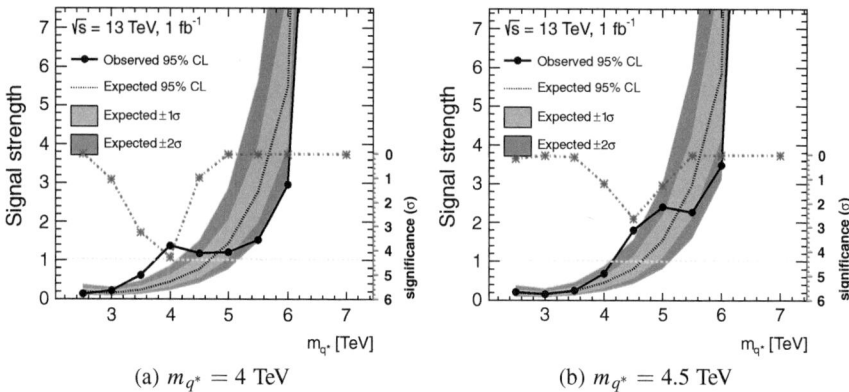

**Fig. 10.7** Sensitivity to injected $q^*$ signal at **a** $m_{q^*} = 4$ TeV and **b** $m_{q^*} = 4.5$ TeV, as function of signal hypothesis $m_{q^*}$. The crossing of the expected limit and the line at signal strength $\mu = 1$ indicates the sensitivity in $m_{q^*}$ of the experiment. The corresponding null hypothesis significance is indicated on the right-hand axis

reach in $m_{q^*}$ stops at around 5 TeV. However, extending the $m_{jj}$ fit reach would not improve the sensitivity much in this example, as the integrated luminosity also plays a role: at 1 fb$^{-1}$, neither the SM nor the signal are expected to produce enough events in the higher $m_{jj}$ range, and including this range would simply contribute empty histograms.

The observed limit in this figure is obtained using pseudo-data drawn from the SM prediction with *injected*[12] signal from the prediction for $m_{q^*} = 4$ TeV in Fig. 10.7a and $m_{q^*} = 4.5$ TeV in Fig. 10.7b. These two values were chosen as they are close to the lower limits on $m_{q^*}$ obtained using Run1 data, and close to the expected limits at 1 fb$^{-1}$. The observed limits visibly deviate from the expected at precisely these values, meaning that only a larger signal than the nominal prediction for these values of $m_{q^*}$ can be excluded—the nominal signal prediction is compatible with the observed distributions (here, by construction). Furthermore, this injected signal resembles the shape of the signal hypothesis at neighbouring $m_{q^*}$, giving a wider bump rather than a spike at the precise signal injection value.[13] Finally, the null hypothesis $p$-value is represented in terms of significance[14] by the orange line, with values to be read off the right-hand axis. In particle physics, a $3\sigma$ significance is often reported in terms of "evidence" of a signal, while $5\sigma$ is an "observation". In this particular example, the experiment would be able to claim discovery of $q^*$ signal at $m_{q^*} = 4$ TeV.[15]

---

[12] Signal injection: adding the signal prediction to the SM prediction.

[13] Remember: the $m_{jj}$ binning is 300 GeV wide.

[14] The $p$-value relates to this $\sigma$ via the standard deviations of a Gaussian distribution: the $p$ value is the probability to obtain a value $q \geq q_0$ where $q_0$ is the observed value, when sampling a Gaussian distribution $G(\mu, \sigma)$. The location of $q_0$ in $G$ is indicated by the number of $\sigma$.

[15] In practice, this is not the wording that would be used, but something more model-agnostic.

## 10.5 Systematic Uncertainties

This section describes how the systematic uncertainties, entering as nuisance parameters in the statistical analysis of data, are obtained. The main principle is to find a baseline $(\vartheta, \sigma_\vartheta)$ for the construction of the Gaussian likelihood for each nuisance parameter, for the number of interest in the analysis (such as the number of entries in a bin). While $\vartheta$ is the value of the nominal prediction, the width of the Gaussian is given by the best knowledge of the impact of a parameter on some variable of interest, for instance the $N_{PV}$ dependence of jet $p_T$, in the assessment of a correction or calibration. It is typically derived from the degree of data/MC disagreement, or MC non-closure,[16] or in the worst case, by the statistical uncertainty in the assessment of the method. Let's call this width $\sigma_{CP}$.[17] The method to find $\sigma_\vartheta$ is to vary each parameter by $\pm 1\sigma_{CP}$ to find the resulting variation in the distribution of interest. Since the angular distribution predictions are normalised to match the data integral, so are the varied predictions.[18] This means that only the impact on the shape of the distribution enters. This is advantageous since for instance the absolute cross section prediction of MC can depend on non-perturbative parameters that need to be tuned, while it's in the nature of this type of high-$p_T$ regime search at the energy frontier that it hasn't been explored much before.

The experimental uncertainties are the JES and luminosity uncertainty, while the theoretical uncertainties relate to the choice of PDF, renormalisation and factorisation scale, MC tuning and generator choice. For the PDF and scale choice uncertainties, NLOJET++ is used together with APPLgrid [10], which lays out a phase space grid for reweighting an input NLO cross section according to different PDF sets and scale settings.

### 10.5.1  JES

For a general jet measurement, the jet energy scale uncertainty introduced in Sect. 6.3 tends to be the dominant uncertainty, and much work goes into reducing it (the reduced pile-up uncertainty from the introduction of the method described in Chap. 7 is one example). The measured angular difference between two jets is not strongly affected by the energy scale, since the jet axis doesn't change. However, the binning in $m_{jj}$ introduces migrations as the $p_T$ shifts (cf. Eq. 8.13).

The first 2015 data JES uncertainty was based on the calibration for 2012 data taking, with a cross-calibration term for the changed conditions (mainly changed

---

[16]*Closure* in a MC- derived correction means independence of the used variables by construction.

[17]The calibrations and corresponding uncertainties are worked out in *Combined Performance* groups.

[18]The systematic variations are a what-if-scenario, implying that they need to be treated in the same manner as the nominal prediction.

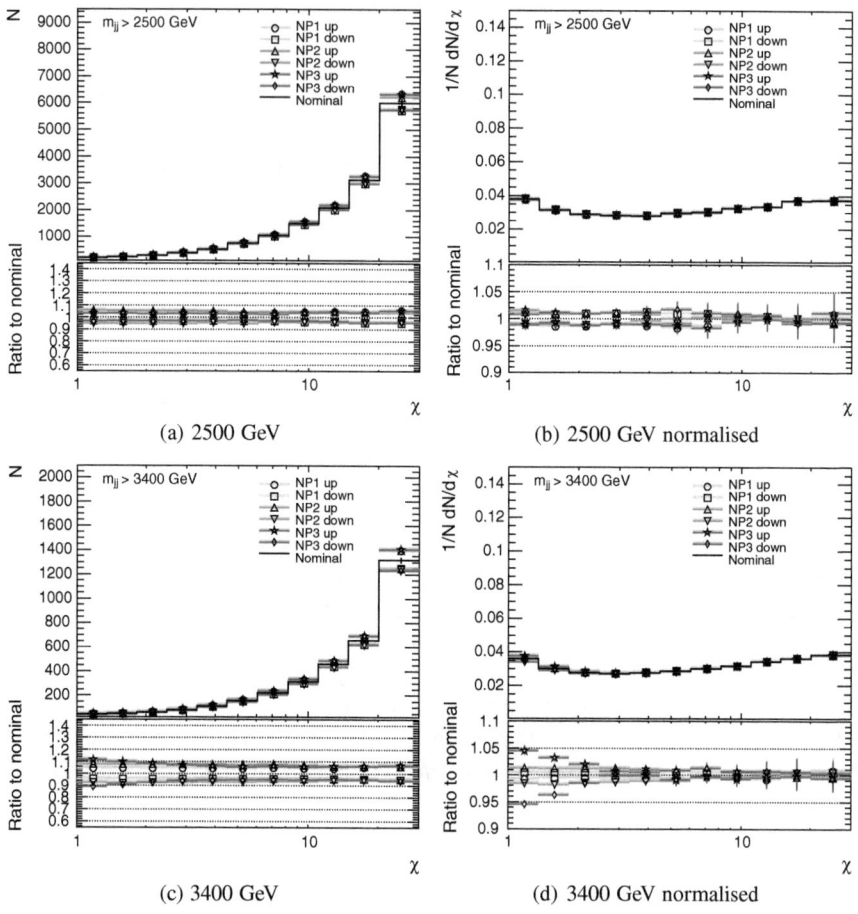

**Fig. 10.8** JES uncertainty breakdown into reduced nuisance parameters 1–3, for **a, b** $2500 \leq m_{jj} < 2800$ GeV and **c, d** $3400 \leq m_{jj} < 3700$ GeV. Figures **a** and **c** show unnormalised $N$ versus $\chi$ distributions, and **b** and **d** show normalised angular distributions. Note the different vertical scales on the ratio panels

bunch spacing and material change due to detector upgrade) [11]. As can be seen in Fig. 6.4a, the cross calibration term becomes negligible at $p_T \sim 300$ GeV.

Figure 10.8 shows the impact on the angular distribution in different $m_{jj}$ regions from varying the JES according to a reduced nuisance parameter (NP) set, diagonalising the 67 components of the JES into three orthogonal NPs [12]. This is done in three different scenarios, expected to be equivalent, unless the phase space of the analysis is sensitive to correlations between the underlying components. The analysis results were indeed found to independent of the choice of reduction scenario.

Comparing Fig. 10.8a–d, what differs is that Fig. 10.8a and c show the $N$ vs $\chi$ distribution, while Fig. 10.8b and d show the normalised differential angular distribution. Differentiating does not affect the ratio, while normalisation does, as every distribution is normalised to unit area. Note that the vertical scales differ between the two versions. It is clear that the final uncertainty is reduced in the normalisation, which only preserves shape differences. The NP3 term dominates in Fig. 10.8d, and this holds also at higher $m_{jj}$. This is the NP set which contains the high-$p_T$ term of the uncertainty, which as we have seen starts to become large at $p_T \sim 2$ TeV and is flat from there on. Here we see that an uncertainty on the jet $p_T$ translates into an uncertainty on $\chi$, owing to $m_{jj}$ migrations.

### $\sqrt{s} = 8$ TeV

In the 2012 JES, a reduced parameter scheme of 14 parameters was used. In this analysis, the impact on the shape of the angular distributions was captured by the $\eta$ intercalibration uncertainty term. The remaining 13 parameters were thus combined in a single term, and the two were used as independent systematic uncertainties.

## 10.5.2  Luminosity Uncertainty

A luminosity uncertainty of $\pm 9\%$ is taken into account for the QBH and $q^*$ signal, used in the analysis of $\sqrt{s} = 13$ TeV data only. For the SM and CI prediction, the normalisation to the data integral removes the uncertainty on the integrated luminosity.

## 10.5.3  PDF Uncertainty

The PDF choice mainly affects the cross section seen in a given $m_{jj}$ region. Thus, even though the PDF uncertainties[19] are generally large at the largest $x$, probed by high-$p_T$ jets, the effect on the normalised angular distributions is very small. This is illustrated in Fig. 10.9, which shows both cases. The uncertainty is calculated using inter- and intra-variations from three different PDF sets, where one is the baseline PDF used in the SM prediction.

---

[19]The details of the PDF uncertainty calculation are given in Appendix A.

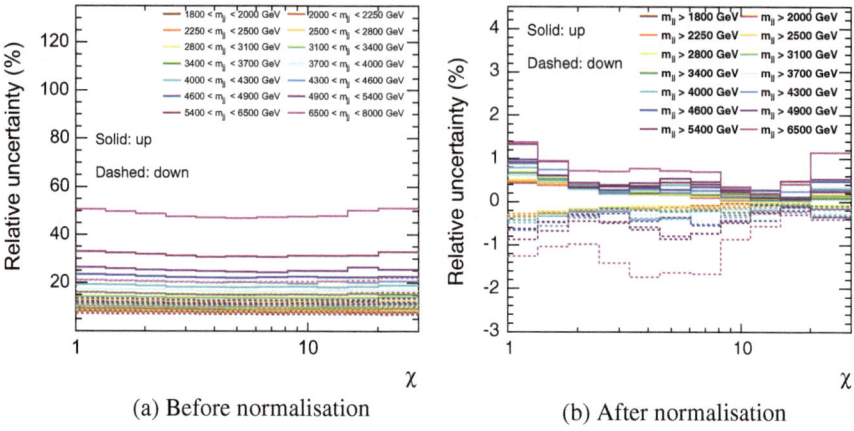

(a) Before normalisation          (b) After normalisation

**Fig. 10.9** Relative PDF uncertainties in **a** unnormalised and **b** normalised angular distributions, for the analysis of $\sqrt{s} = 13$ TeV data

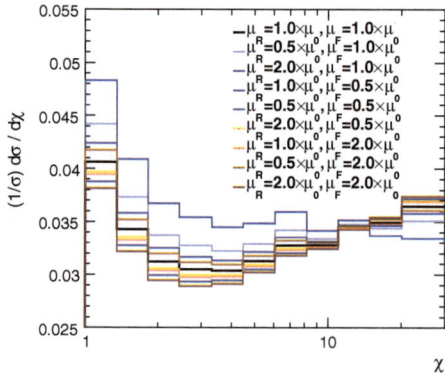

**Fig. 10.10** Angular distributions with scale variations, for $3700 \leq m_{jj} < 4000$ GeV, evaluated at $\sqrt{s} = 13$ TeV

## 10.5.4  Scale Uncertainty

The scales for renormalisation and factorisation are typically set at $Q$, which is not exactly experimentally measured. As a proxy, $\mu_R = \mu_F = \frac{p_T^{lead} + p_T^{sublead}}{2}$ is chosen. To assess the uncertainty from this choice, $\mu_R$ and $\mu_F$ are independently varied up and down by a factor of 2 (which is an arbitrary but conventional choice). The resulting distributions are exemplified for one region of $m_{jj}$ in Fig. 10.10.

In the analysis of $\sqrt{s} = 8$ TeV data, the $1\sigma$ uncertainty on the scale choice was taken as the RMS of the distributions thus obtained, for each $m_{jj}$ region. In the analysis of 13 TeV data, the $1\sigma$ uncertainty was instead taken as the envelope of the resulting distributions with anti-variations excluded. In both cases, the final uncertainty is

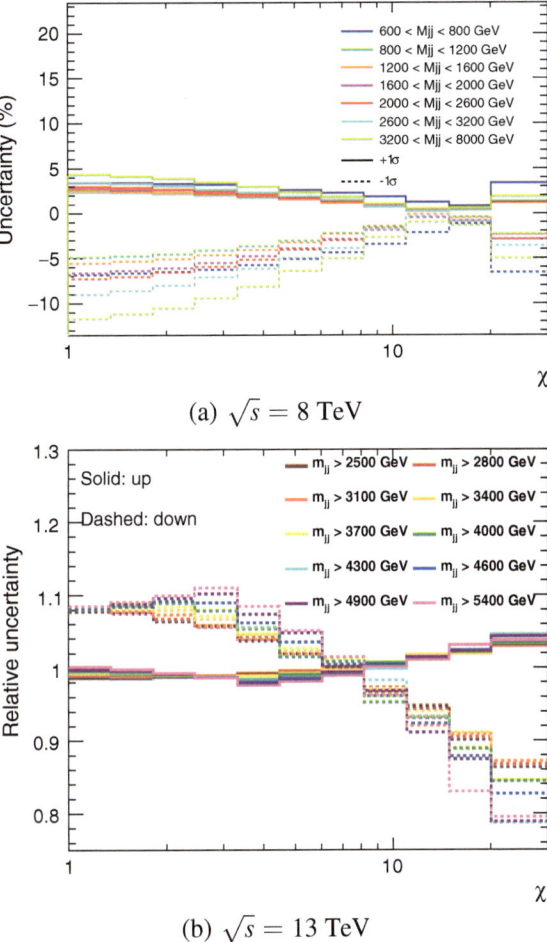

**Fig. 10.11** Relative scale uncertainties in normalised angular distributions, for several $m_{jj}$ regions, for **a** $\sqrt{s} = 8$ TeV and **b** $\sqrt{s} = 13$ TeV

assessed on distributions normalised as a last step. Figure 10.11 shows the resulting scale uncertainty from the two approaches. The shape differs, but also the magnitude, with the procedure used in the analysis of $\sqrt{s} = 13$ TeV data (Fig. 10.11b) giving larger uncertainty.

## 10.5.5 Tune Uncertainty

The uncertainty due to the tuning choice of parameters governing non-perturbative as well as perturbative processes is assessed by variations. The non-perturbative processes are not expected to affect high-$p_T$ jets very much.

**8 TeV**

No clear evidence of improved tune settings were seen in the studies outlined in Sect. 9.1. Instead the envelope of the distributions was used to calculate an asymmetric tune uncertainty. Data and MC agreed within this uncertainty at low $m_{jj}$. At $\sim 2$ TeV, the different tunes all converged to the baseline prediction, modulo statistical fluctuations, which were non-negligible.

Similarly, a generator uncertainty was derived by comparing the prediction of POWHEG showered with PYTHIA8.175 to PYTHIA8 brought to NLO with $K$-factors, and a showering uncertainty was obtained from the comparison of POWHEG+PYTHIA8 to POWHEG showered with HERWIG+JIMMY [13–17], v6.520.2 and v4.31. These were both dominated by statistical uncertainties.

**13 TeV**

A central, large-statistics production of particle-level MC samples with Professor eigentune variations [18, 19] was used to obtain the envelopes of the varied distributions. The same smoothing procedure used for the $K$-factors was applied to remove an otherwise large impact of statistical uncertainty from points insignificantly deviating from the nominal prediction. The resulting tune uncertainty is shown before and after normalisation in Fig. 10.12. While the tune choice does affect the overall cross section, it does not give a strong angular shape dependence[20] in the normalised distributions and is negligible compared to the JES and scale uncertainty.

## 10.6  Total Uncertainty

Examples of the total uncertainty in the regions used for statistical analysis, and a breakdown into the above components is shown below. Figure 10.13 shows all the evaluated components in the $\sqrt{s} = 8$ TeV analysis, for $m_{jj} > 3200$ GeV. Upward and downward variations are shown as solid and dashed lines, respectively.[21] It is clear that the JES and scale uncertainties dominate. The PDF and $K$-factor uncertainties are hardly discernible.

For the $\sqrt{s} = 13$ TeV analysis in Figs. 10.14a–c, only the two major ones are shown. The scale uncertainty is large at $m_{jj} = 3400$ GeV, while for $m_{jj} > 5400$ GeV, the JES uncertainty is much larger. The JES uncertainty is more symmetric than the scale uncertainty. Furthermore, at high $m_{jj}$ it is flatter in $\chi$, which is attributed to the high-$p_T$ term of the uncertainty: this is flat in $p_T$ from 2 TeV up, and starts making its way in from the low $\chi$ region, following Eq. 8.13.

---

[20]Shape effects are, as we have seen, a sign of both angular shifts and $m_{jj}$ migrations.

[21]The generator and shower uncertainties are one-sided.

**Fig. 10.12** Relative tune
uncertainties in the
$\sqrt{s} = 13$ TeV angular
distributions, for several $m_{jj}$
regions, **a** before and **b** after
normalisation

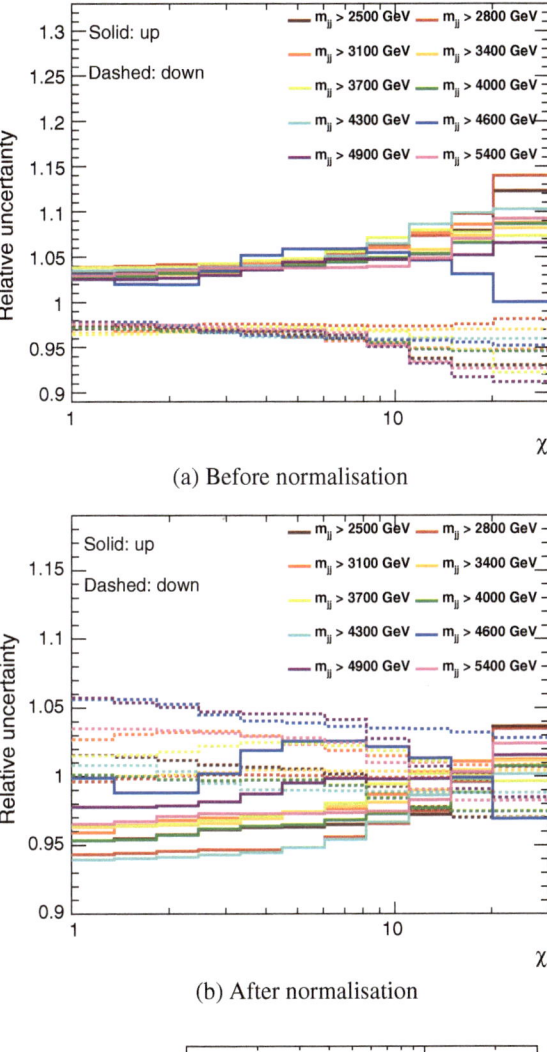

(a) Before normalisation

(b) After normalisation

**Fig. 10.13** Breakdown of
the total uncertainty in the
$\sqrt{s} = 8$ TeV analysis, for the
limit setting region
$m_{jj} > 3200$ GeV. Upward
and downward variations are
drawn with solid and dashed
lines, respectively

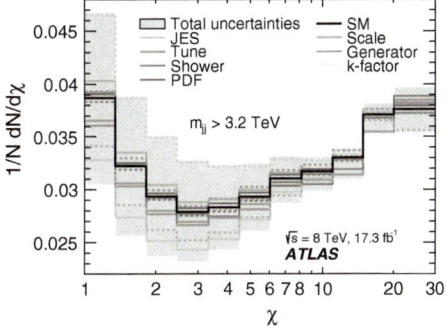

**Fig. 10.14** Breakdown of the total uncertainty in the $\sqrt{s} = 13$ TeV analysis, for **a** $3400 \leq m_{jj} < 4000$ GeV, **a** $m_{jj} > 3400$ GeV and **c** $m_{jj} > 5400$ GeV. Upward and downward variations are drawn with solid and dashed lines, respectively

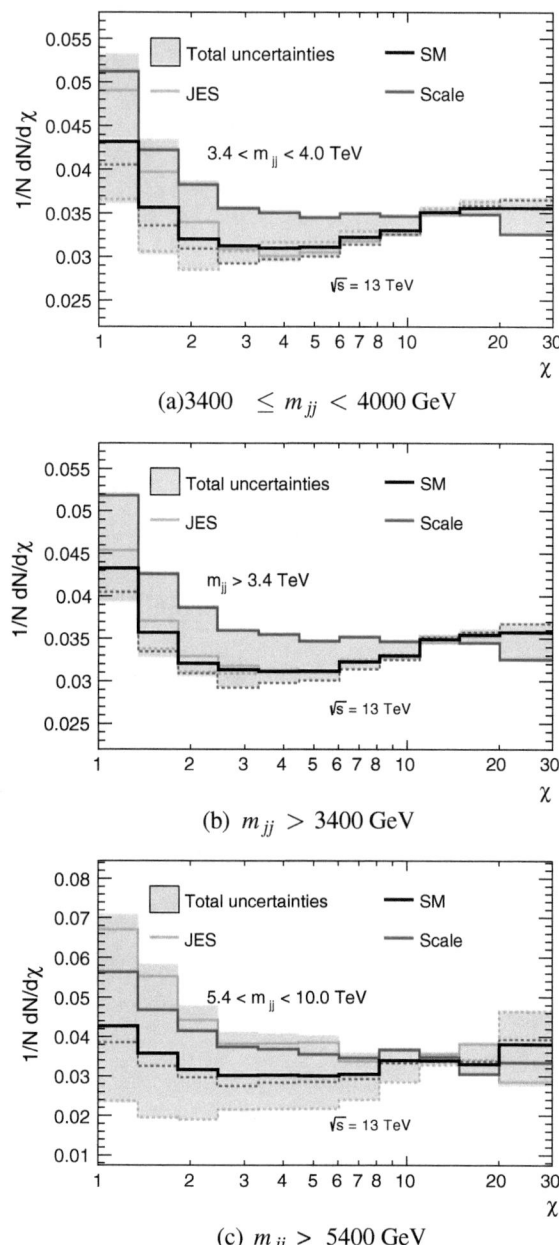

(a) $3400 \leq m_{jj} < 4000$ GeV

(b) $m_{jj} > 3400$ GeV

(c) $m_{jj} > 5400$ GeV

# References

1. ATLAS Collaboration, Search for new phenomena in dijet angular distributions in proton-proton collisions at $\sqrt{s}$ = 8 TeV measured with the ATLAS detector. Phys. Rev. Lett. **114**, 221802 (2015)
2. Search for New Phenomena in Dijet Mass and Angular Distributions with the ATLAS Detector at $\sqrt{s}$ = 13 TeV. Technical Report, ATLAS-CONF- 2015-042. Geneva: CERN (2015)
3. ATLAS Collaboration, Search for new phenomena in dijet mass and angular distributions from pp collisions at $\sqrt{s}$ = 13 TeV with the ATLAS detector. Phys. Lett. B **754**, 302–322 (2016). ISSN: 0370-2693
4. W.J. Stirling, Private Communication
5. V. Lendermann et al., Combining triggers in HEP data analysis (2009), arXiv:0901.4118
6. J.H Friedman, Data analysis techniques for high energy particle physics, SLAC-176 (1974), p. 96
7. S. Dittmaier, A. Huss, C. Speckner, Weak radiative corrections to dijet production at hadron colliders. JHEP **1211**, 095 (2012)
8. K. Cranmer et al., HistFactory: a tool for creating statistical models for use with RooFit and RooStats. CERN-OPEN-2012-016 (2012)
9. L. Moneta et al., The RooStats project, in *Proceedings of the 13th International Workshop on Advanced Computing and Analysis Techniques in Physics Research* (2010), p. 57, arXiv:1009.1003
10. Tancredi Carli et al., A posteriori inclusion of parton density functions in NLO QCD final-state calculations at hadron colliders: the APPLGRID project. Eur. Phys. J. C **66**, 503–524 (2010)
11. ATLAS Collaboration, Jet calibration and systematic uncertainties for jets reconstructed in the ATLAS detector at $\sqrt{s}$ = 13 TeV. ATLPHYS- PUB-2015-015 (2015)
12. ATLAS Collaboration, A method for the construction of strongly reduced representations of ATLAS experimental uncertainties and the application thereof to the jet energy scale. Technical Report, ATL-PHYSPUB- 2015-014. Geneva: CERN (2015)
13. G. Corcella et al., HERWIG 6.5 release note (2002), arXiv:hep-ph/0210213
14. G. Marchesini et al., A Monte Carlo event generator for simulating hadron emission reactions with interfering gluons. Comput. Phys. Commun. **67**, 465–508 (1991)
15. G. Marchesini et al., Monte Carlo simulation of general hard processes with coherent QCD radiation. Nucl. Phys. B **310**, 461 (1988)
16. B.R. Webber, A QCD model for jet fragmentation including soft gluon interference. Nucl. Phys. B **238**, 492 (1984)
17. J.M. Butterworth, J.R. Forshaw, M.H. Seymour, Multiparton interactions in photoproduction at HERA. Z. Phys. C **72**, 637–646 (1996)
18. ATLAS Collaboration, ATLAS Run 1 Pythia8 tunes. Technical Report, ATL-PHYS-PUB-2014-021. Geneva: CERN (2014)
19. ATLAS Physics Modelling Group, PMG recommendations for tune uncertainty. ATLAS internal (2015), https://twiki.cern.ch/twiki/bin/view/AtlasProtected/MCTuningRecommendations

# Chapter 11
# Results

In this chapter, we have finally reached the goal: all the pieces are in place to have a look at and interpret the physics message in the dijet angular distributions. Here I show the $\sqrt{s} = 8$ and 13 TeV results, discuss and compare them, and discuss some possible paths forward.

## 11.1 Angular and Mass Distributions

### 11.1.1 8 TeV

Figure 11.1 shows the normalised angular distributions in the $\sqrt{s} = 8$ TeV data, overlaid with the MC SM prediction with and without EW corrections applied. Theoretical uncertainties are shown as a shaded band. Experimental uncertainties (meaning, the JES uncertainty) are shown as a dash on the vertical error bars, which represent the statistical and experimental uncertainties added in quadrature. The predicted CI signal for two combinations of $\Lambda$ and $\eta_{LL}$ is also shown.

It is clear that the EW corrections, used here for the first time in an ATLAS dijet search, improve the data/MC agreement significantly in the region of low $\chi$, high $m_{jj}$. This is precisely the region where one would naïvely expect new phenomena to occur first; for instance this is the most sensitive region to the CI signal. This correction does bring a significant improvement also to the limits on new phenomena in the absence of significant deviations from the SM. Another clear feature is the data/MC discrepancy in the low $m_{jj}$ region, covered by the tune uncertainty, and closing in the region where all tunes converge (around 2 TeV). The theoretical uncertainties dominate at low $m_{jj}$; the JES uncertainty grows with $m_{jj}$, as does the statistical uncertainty, which is only noticeable in the last $m_{jj}$ window.

Since all distributions are normalised to unit area, comparing the shape across $m_{jj}$ regions is easily done. Figure 11.2 shows a comparison of the distribution in each $m_{jj}$ region to the weighted average over all the rest. This way of averaging

© Springer International Publishing AG 2017
L.K. Bryngemark, *Search for New Phenomena in Dijet Angular Distributions at $\sqrt{s}$ = 8 and 13 TeV*, Springer Theses, DOI 10.1007/978-3-319-67346-2_11

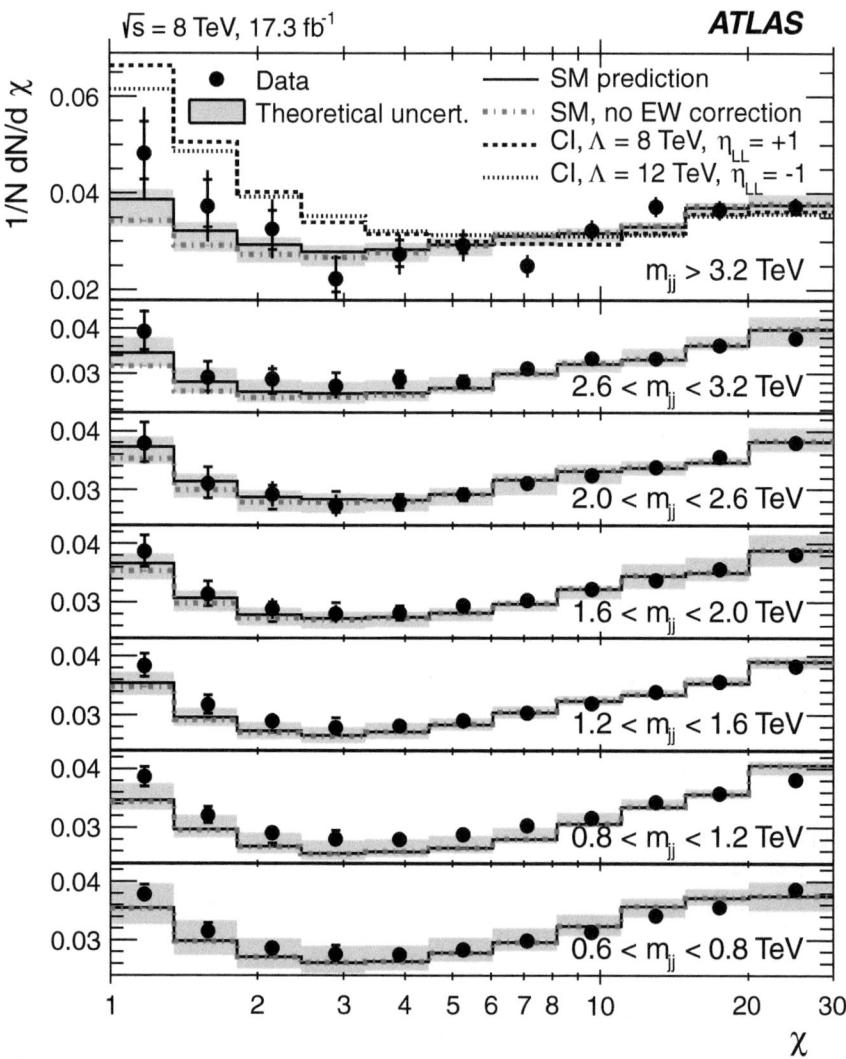

**Fig. 11.1** Normalised angular distributions in the $\sqrt{s} = 8$ TeV data, overlaid with the MC prediction with and without EW corrections applied. Theoretical uncertainties are shown as a shaded band. Experimental uncertainties are shown as a dash on the vertical error bars, which represent the statistical and experimental uncertainties added in quadrature. The predicted CI signal for two choices of $\Lambda$ and $\eta_{LL}$ is also shown

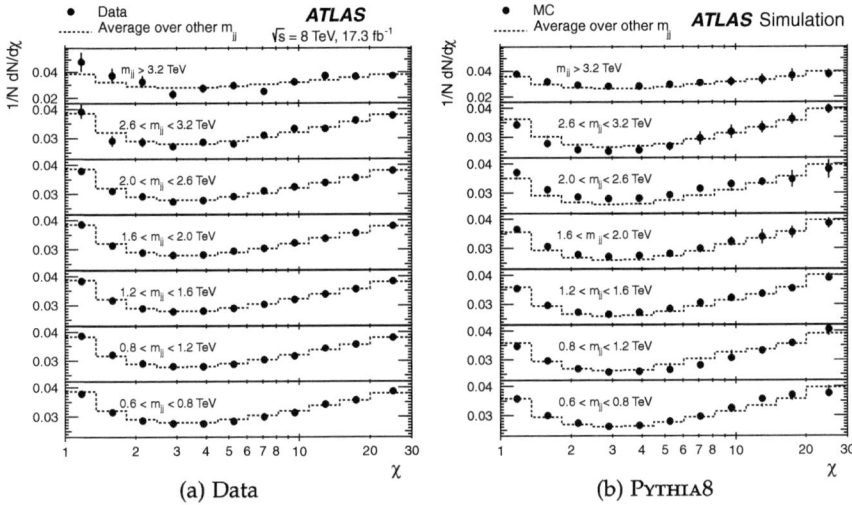

**Fig. 11.2** Comparison of the normalised angular distribution in each $m_{jj}$ region to all the others, except the highest, in **a** data and **b** NLO QCD and EW corrected PYTHIA8

**Fig. 11.3** $F_\chi$ as derived from the angular distributions. Comparison of $\sqrt{s} = 8$ TeV data, pure PYTHIA8 and fully corrected PYTHIA8

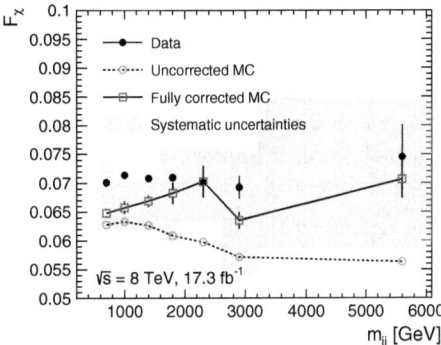

means that the lowest $m_{jj}$ distribution will dominate the comparison. It is striking in Fig. 11.2a that all $m_{jj}$ regions in data except the last one agree very well with the average, meaning, that this dominance from the low $m_{jj}$ doesn't matter much. In MC, shown in Fig. 11.2b, there are larger variations, but no general trend.[1] This variation is attributed to statistical fluctuations.

The impact of the corrections on the shape can also be visualised in the $F_\chi$ distribution. Neither final corrections nor systematic uncertainties were derived for this distribution. Since $F_\chi$ corresponds to the ratio of the number of events in the first four bins of the angular distributions to the total number of events, it can be recovered from the angular distributions, albeit in the coarse $m_{jj}$ binning.

---

[1] Without EW corrections, there is a clear systematic trend with $m_{jj}$ in the shapes.

**Fig. 11.4** Dijet mass
distribution in the $\sqrt{s} = 8$
TeV data, compared to the fit
(red line) in the bottom
panel, with signal prediction
for three $m_{q^*}$ overlaid (color
figure online)

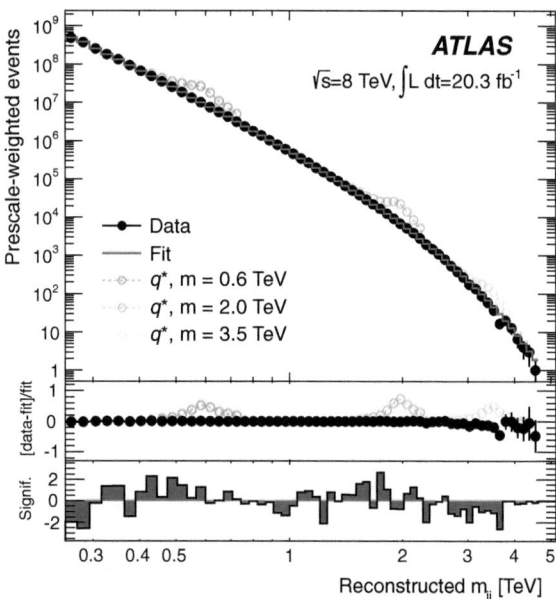

Figure 11.3 shows $F_\chi$ in data and MC, where both the pure PYTHIA8 prediction
and the fully corrected SM prediction is shown. Total systematic uncertainties are
included. It was verified for $\sqrt{s} = 8$ TeV data and uncorrected MC that the original
$F_\chi$ finely binned in $m_{jj}$ gives the same result as the one derived from the angular
distributions.

Since $F_\chi$ and the angular distributions carry the same information, Fig. 11.3 again
shows that there is a difference in the shape of the angular distributions in data and
MC, which decreases as $m_{jj}$ increases.[2] This is the region where the $K$- and $\kappa$-factors
play the largest role. At low $m_{jj}$, as we have seen in the angular distributions, the
uncertainty due to the choice of tune closes the gap between data and SM prediction.
The high degree of $m_{jj}$ independence in data is again obvious.

The dijet mass distribution is shown in Fig. 11.4.

In addition to the data, the fit obtained with a smooth functional form,

$$f(x) = p_1(1 - x)^{p_2} x^{p_3 + p_4 \ln x + p_5 (\ln x)^2}, \qquad (11.1)$$

is shown. Here $x \equiv m_{jj}/\sqrt{s}$, the $p_i$ are fit parameters and $p_5$ was set to zero. The
ratio panel in the middle shows the relative difference to the fit value in each bin, as
well as the expected deviation in the presence of three different $q^*$ signal hypotheses.
The bottom panel shows the bin-by-bin significance in the comparison of the data
and the fit, taking only statistical uncertainties into account.

---

[2]In the uncorrected MC, the gap instead increases, which raised concerns about the ability to model
$F_\chi$ in MC.

The fit shows large oscillations, which are not present in a fit allowing a non-zero fifth parameter. The fit function choice was based on a blinded analysis, using a partial dataset.[3] With the larger statistical power of the full data set, an additional parameter would have been necessary to accommodate the level of detail resolved in the spectrum.

For the 2015 iteration of the analysis, a smaller data set was expected, covering a smaller range in $m_{jj}$ owing to the use of only un-prescaled triggers. Thus, a lower-order parameterisation was chosen as a starting point, with the fourth and fifth parameter initially set to zero. A pre-defined figure of merit, based on hypothesis testing, was implemented. It uses the $p$-value from a log-likelihood ratio, the lower-order parameterisation being the null hypothesis, and the higher-order parameterisation being the alternate, to find the fit preferred by the data. If the $p$-value drops below 0.05 as more integrated luminosity is added, the lower-order parameterisation is discarded. This procedure automatises the inclusion of higher parameters if needed as the integrated luminosity increases, removing the need for a blinded analysis, and enabling the flexibility missing in the analysis of $\sqrt{s} = 8$ TeV data.

### 11.1.2  13 TeV

Figure 11.5 shows normalised angular distributions in the $\sqrt{s} = 13$ TeV data, overlaid with the fully corrected MC prediction. Theoretical and total uncertainties are shown as lighter and darker shaded bands, while the vertical error bars represent the statistical uncertainty. The predicted CI signal for two choices of $\Lambda$ and $\eta_{LL}$ is also shown, along with a QBH prediction. In particular, the signal is shown for all the $m_{jj}$ regions used in the combined fit, demonstrating the additional power in taking the signal evolution in $m_{jj}$ into account. This will be discussed in more detail in Sect. 11.2.2.

Once again, $m_{jj}$ independence of the normalised angular distributions is observed. This is demonstrated for $\sqrt{s} = 13$ TeV in Fig. 11.6, obtained in the same manner as Fig. 11.2. The variation in MC is no longer present in Fig. 11.6b. This is attributed to smaller fluctuations in the $K$-factors due to larger samples used for their derivation. It is clear that the data and MC distributions are both internally consistent across $m_{jj}$. The data distributions show a small wiggle at intermediate $\chi$ not present in the MC prediction, which has a slightly flatter shape in $\chi$ than data.

The $m_{jj}$ spectrum is shown in Fig. 11.7, together with the 3-parameter fit described by Eq. 11.1 and overlaid signal prediction for $q^*$ and QBH at two mass hypotheses. The $q^*$ signal has been scaled up by a factor 3 for visibility. No higher-order parameterisation was needed to describe the data. The most discrepant range is at $m_{jj} = 1.5-1.6$ TeV, indicated by the two vertical lines, but no significant excess is observed. The middle panel shows the bin-by-bin significance of the deviations in the data from the fit, and the bottom panel shows the comparison of the data to the

---

[3] Also here, 1/4 of the data set was used for analysis optimisation.

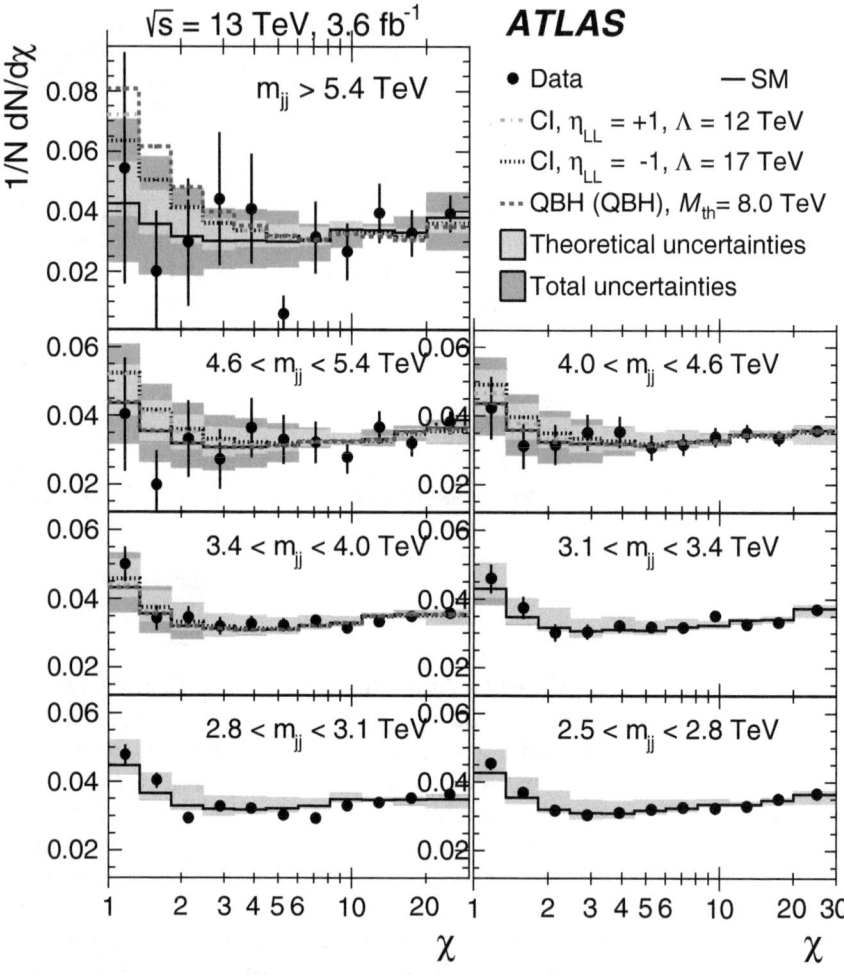

**Fig. 11.5** Normalised angular distributions in the $\sqrt{s} = 13$ TeV data, overlaid with the fully corrected MC prediction. Theoretical and total uncertainties are shown as lighter and darker shaded bands, while the vertical error bars represent the statistical uncertainty. The predicted CI signal for two choices of $\Lambda$ and $\eta_{LL}$ is also shown, along with a QBH prediction

PYTHIA8 prediction, corrected with dedicated $K$- and $\kappa$-factors for the mass analysis event selection, and normalised to match the data integral. The JES uncertainty band is also drawn. Although data agree with the MC prediction within uncertainties, an overall trend can be observed where the shape in data falls off at higher $m_{jj}$. The trend in the agreement of data and MC points to several things. Firstly, we are probing an energy regime that has not been explored before. One should however note that we are far from the most extreme regions of the PDFs here (the dijet mass spectrum at $\sqrt{s} = 8$ TeV is closer to that regime), and it is likely that, with more data, tuning of

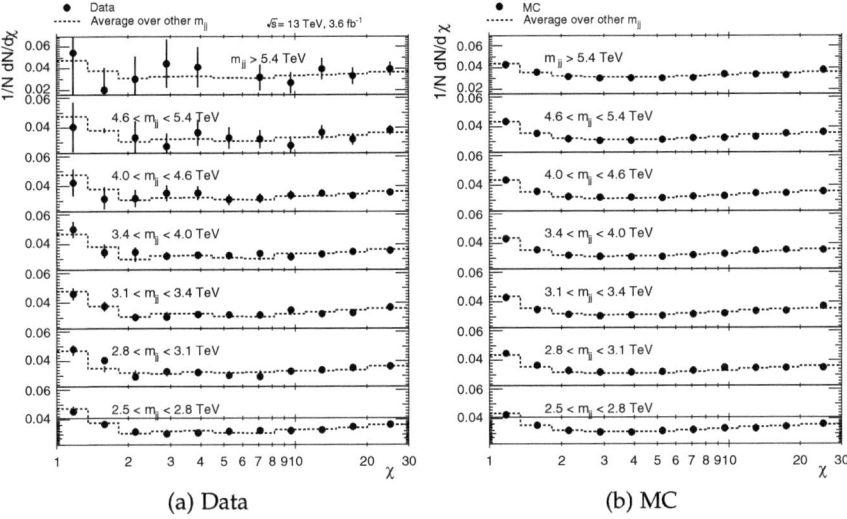

(a) Data          (b) MC

**Fig. 11.6** Comparison at $\sqrt{s} = 13$ TeV of the normalised angular distribution in each $m_{jj}$ region to all the others, except the highest, in **a** data and **b** NLO QCD and EW corrected PYTHIA8

**Fig. 11.7** Dijet mass distribution in the $\sqrt{s} = 13$ TeV data, compared to the fit (red line) in the second panel, with signal prediction for three $m_{q*}$ overlaid. The bottom panel shows the comparison to NLO QCD and EW corrected PYTHIA8, along with the JES uncertainty (color figure online)

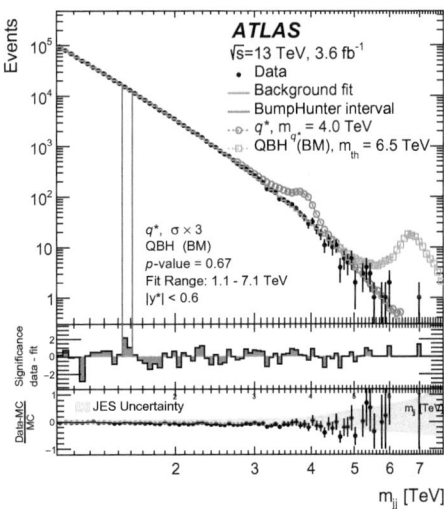

parameters will improve. This comparison is however a striking demonstration that for a "first search" like this one, using a data-driven SM prediction is very advantageous. Secondly, it is clear that the normalisation of MC to the data integral in each

$m_{jj}$ region removes effects like these from the angular distributions, and allows for focusing on the shape.[4]

## 11.2  Statistical Analysis and Limits

We will now proceed to the quantitative analysis of the angular distributions and the implications for theories beyond the SM. Note that when interpreting the angular distribution figures, in particular with respect to normalisation impact or statistical analysis, one always needs to bear in mind that they are shown with a logarithmic horizontal axis and that the statistical power of the high-$\chi$ region is larger.

### 11.2.1  Fit Control Plots, Analysis of $\sqrt{s} = 13\,TeV$ Data

The constraints on a few example nuisance parameters obtained in the fit of the SM prediction to the data are shown in Fig. 11.8. The constraints in each $m_{jj}$ region on the reduced JES NPs 1 and 3, as well as on the scale uncertainty, are shown as a deviation from the nominal $\theta$ and $\sigma$, where $1\sigma$ and $2\sigma$ are indicated by the green and yellow bands.

Figure 11.8a shows that there is not enough statistical power in this data set to constrain the JES NP1 uncertainty, since this is very small. The JES NP3, on the other hand, gets pulled and constrained in the fit across all $m_{jj}$, as seen in Fig. 11.8b. This reduced NP contains the high-$p_{\mathrm{T}}$ uncertainty term, which is the dominant term in the JES uncertainty.[5] We see that the data suggest a smaller high-$p_{\mathrm{T}}$ uncertainty. Finally, the scale uncertainty shown in Fig. 11.8c gets more constrained at lower $m_{jj}$, where it is clearly dominant, and less so at higher $m_{jj}$, where the JES uncertainty starts to dominate (cf. the $m_{jj}$ evolution in Fig. 10.14).

To exemplify the interplay of the size of the statistical and systematic uncertainty, the shape of the scale and JES NP3 uncertainties in the lowest and highest $m_{jj}$ region is shown in Fig. 11.9. It is clear in Fig. 11.9b that the statistical precision in the data is not enough to constrain the scale uncertainty at high $m_{jj}$, while from the other three figures, one can expect some constraint on the nuisance parameters from the fit to data.

---

[4] An equivalent approach is to leave the normalisation as a free parameter in the fit. Once the overall normalisation is better modelled, fitting through varied nuisance parameters is even a tuning of sorts.

[5] This has been verified in tests using the 3 other reduced NP scenarios. In all cases, the reduced NP containing the high-$p_{\mathrm{T}}$ term is dominant, and gets constrained similarly.

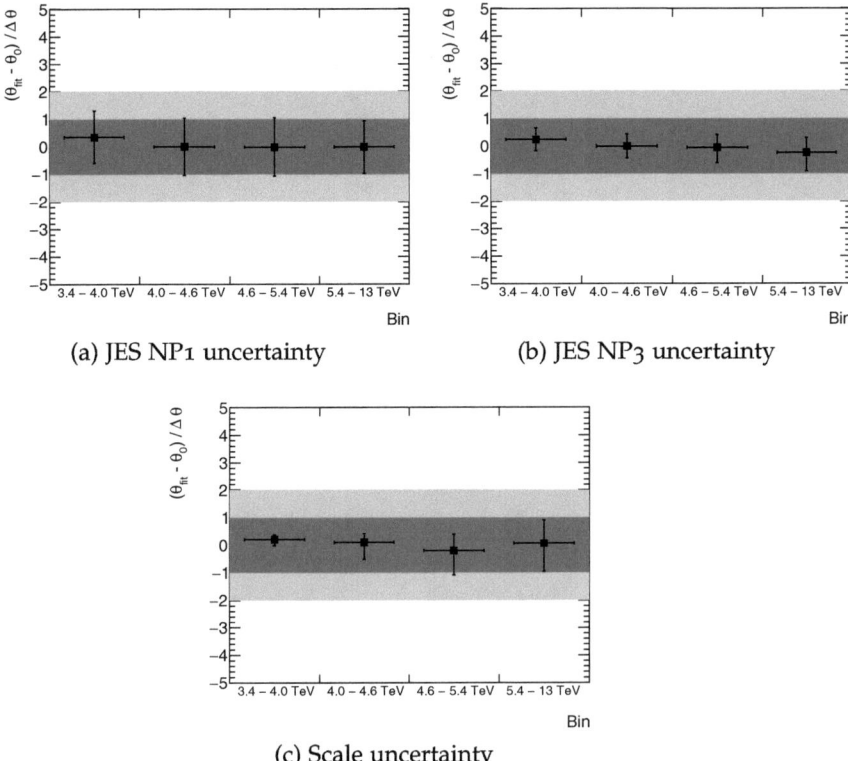

**Fig. 11.8** Constraints on nuisance parameters **a** JES, NP1, **b** JES, NP3 and **c** scale uncertainty, in the profile likelihood fit, shown for the different $m_{jj}$ regions

## 11.2.2  Limits on the Scale of New Phenomena

The $\sqrt{s} = 8$ TeV data distributions analysed are in excellent agreement with the SM prediction in the $2.0 \leq m_{jj} < 2.6$ TeV region, with a null hypothesis $p$-value of 0.25 and 0.30, respectively, in the $2.6 \leq m_{jj} < 3.2$ and $m_{jj} \geq 3.2$ TeV regions. In the $\sqrt{s} = 13$ TeV data, the null hypothesis $p$-value is 0.35, obtained for the combined fit in the whole region $m_{jj} > 3.4$ TeV.

In the absence of significant deviations between the data and the SM predictions, limits on parameters of benchmark models can be derived from the level of agreement between data and MC, and the shape of any deviations.

$\sqrt{s} = 8$ *TeV: CI* $\Lambda$

The limits on $\Lambda$ in the two modes of interference between CI and QCD modelled are shown in Fig. 11.10. The lower limits are placed at $\Lambda = 8.1$ and $12.0$ TeV for destructive ($\eta_{LL} = +1$) and constructive ($\eta_{LL} = -1$) interference, respectively, the

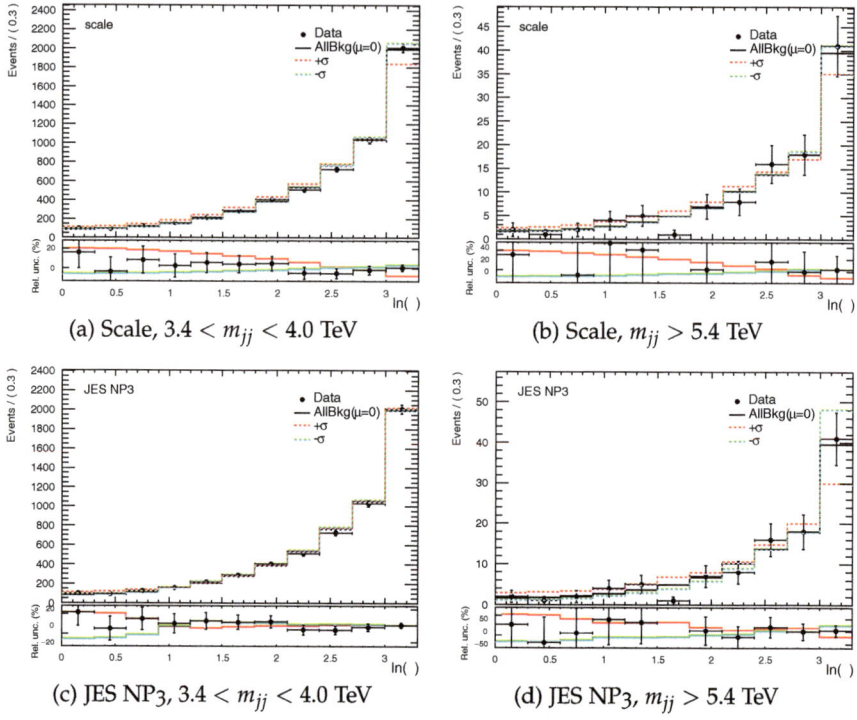

**Fig. 11.9** The uncertainty from the choice of renormalisation and factorisation scales overlaid with the data, for $m_{jj}$ regions **a** $3.4 < m_{jj} < 4.0$ TeV and **b** $5.4 < m_{jj} < 13$ TeV, and for the same two regions for JES NP3 in **c** and **d**, respectively

latter being the most stringent limit on constructive interference set by the end of Run1.

Since only one $m_{jj}$ region is used, the observed limit is given by comparing one and the same shape in data to the shapes from adding in signal at varied signal strength. While the expected limit is given by systematic and expected statistical uncertainties only, the observed limit is based on the actual outcome in data. Data show[6] an insignificant upward deviation at low $\chi$ in the used $m_{jj}$ region, resulting in weaker observed than expected limits. The agreement between the expected and observed limits in the two cases is very different, but internally mostly consistent across $\Lambda$. Especially for constructive interference, the evolution with $\Lambda$ is smooth: the resulting signal contribution quickly gets dominated by the interference term, which is proportional to $\frac{1}{\Lambda^2}$. This means that the shape between the different predictions scaled up by the needed signal strength will not vary much with $\Lambda$, and the same relation between the expected and observed limit holds. For destructive interference,

---

[6]See Fig. 11.1, the $m_{jj} > 3.2$ TeV region.

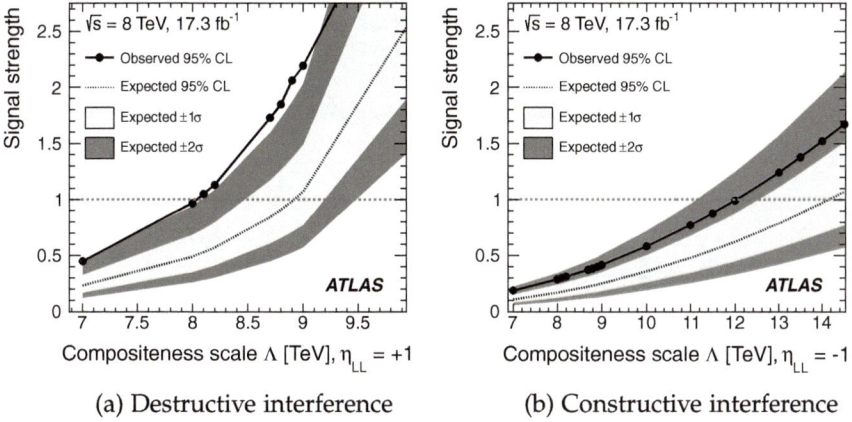

(a) Destructive interference       (b) Constructive interference

**Fig. 11.10** Lower limits derived from the $\sqrt{s} = 8$ TeV data analysis on $\Lambda$ with **a** destructive ($\eta_{LL} = +1$) and **b** constructive ($\eta_{LL} = -1$) interference with QCD

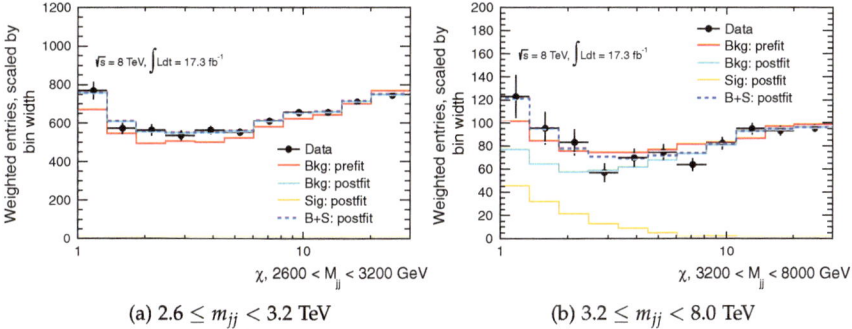

(a) $2.6 \leq m_{jj} < 3.2$ TeV       (b) $3.2 \leq m_{jj} < 8.0$ TeV

**Fig. 11.11** The SM and destructive interference CI signal prediction before and after the profile likelihood ratio fit, for **a** $2.6 \leq m_{jj} < 3.2$ TeV and **b** $3.2 \leq m_{jj} < 8.0$ TeV

at some $\Lambda$ the signal manifests itself as a deficit,[7] implying that the reach in $\Lambda$ is always shorter for destructive interference. The shape of the signal is steeper in $\chi$ in the destructive case than the constructive.

Some insight in the shape of the observed limit for destructive interference in Fig. 11.10a can be gained from comparing the SM and signal predictions before and after the fit to the shape observed in data, as shown in Fig. 11.11. We see in Fig. 11.11a that for a smooth discrepancy between data and SM prediction, that follows the shape of a nuisance parameter, no signal is needed to obtain a good fit. In the case of a steeper excess, followed by a deficit as in the highest $m_{jj}$ region of Fig. 11.1, a non-zero signal strength achieves a better fit than the prediction from SM only, as shown

---

[7] In the region where QCD dominates but still overlaps with signal contributions, larger cancellation occurs.

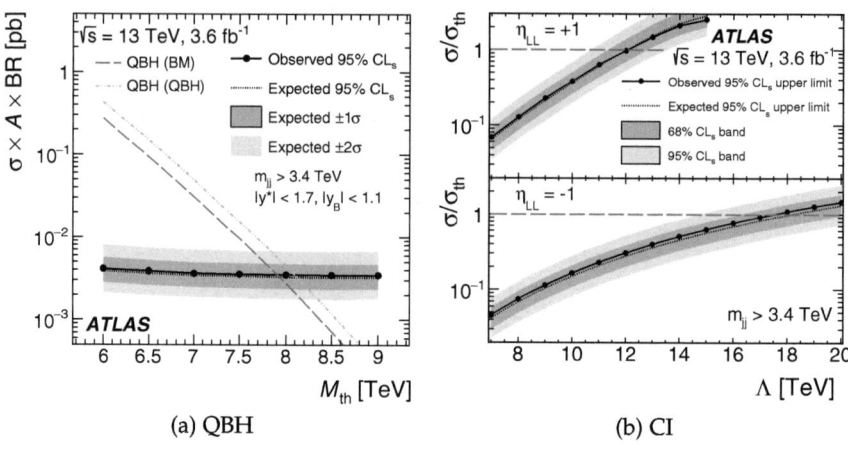

**Fig. 11.12** Lower limits on the scale of new phenomena derived from the $\sqrt{s} = 13$ TeV data analysis: **a** on QBH $M_{th}$ as generated by the QBH (QBH) and BlackMax (BM) generators and **b** limits on CI $\Lambda$ with destructive ($\eta_{LL} = +1$) and constructive ($\eta_{LL} = -1$) interference with QCD

in Fig. 11.11b. Here, especially the steep shape in $\chi$ of the destructive interference signal can easily fill the resulting gap.

### $\sqrt{s} = 13$ TeV: CI $\Lambda$, QBH $M_{th}$ and $m_{q*}$

The limits expressed in $\sigma \times A \times$ BR on $M_{th}$ in the predictions of BlackMax and QBH are shown in Fig. 11.12a. The sensitivity to QBH of angular distributions is, somewhat surprisingly, comparable to the sensitivity in the mass distribution, which has much smaller systematic uncertainties. Given the coarse binning,[8] especially in the higher $M_{th}$ region probed by these data, it is this signal's large cross section that makes it discernible in the angular distributions.

The limits on CI $\Lambda$ are shown in Fig. 11.12b, with the same symbol conventions as before, but denoting signal strength with $\sigma/\sigma_{th}$, as a fraction of the theoretical cross section prediction. In the destructive signal case, there is an evolution in the agreement between observed and expected limit for different $\Lambda$, while for constructive interference there is a constant shift. At low $\Lambda$ the interference term matters less, and the two signal shapes are more similar than at higher $\Lambda$. On top of the impact of the shape difference in a single $m_{jj}$ region as described above, the combined fit introduces sensitivity to the evolution in $m_{jj}$, which is also different in the two interference modes, as seen in Fig. 11.5. Destructive interference CI grows more rapidly with $m_{jj}$ than the constructive one: while destructive CI at the observed limit has zero signal at lower $m_{jj}$ (3.4 TeV), and constructive is non-zero, the destructive interference signal is larger at high $m_{jj}$ (5.4 TeV) than the constructive interference signal at the observed limit. This is not too surprising: when there is much QCD, destructive and constructive interference will result in less/more signal respectively,

---

[8]The highest $m_{jj}$ region starts from 5.4 TeV while an 8 TeV QBH signal peaks at just above 8 TeV.

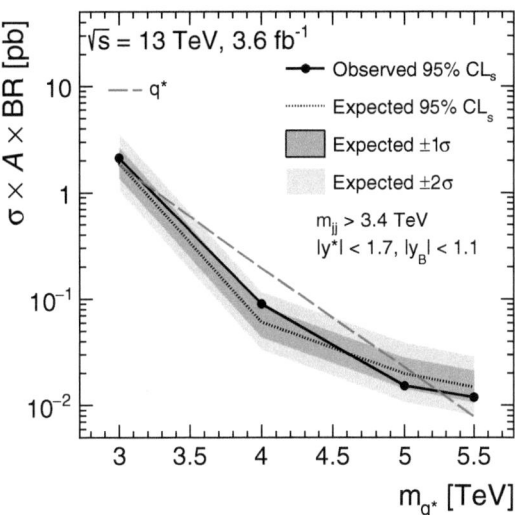

**Fig. 11.13** Lower limits on $m_{q^*}$ derived from the angular distributions in $\sqrt{s} = 13$ TeV data

while as QCD dies off, the interference term makes less difference. Thus, the combined fit retains more information than the single bin fit.[9] This is also seen from the fact that the signal predictions in the highest $m_{jj}$ region of Fig. 11.1 are more similar than in Fig. 11.5, despite the fact that all of them illustrate the signal prediction at the observed limit. Finally, the shift in the observed destructive interference limit from being weaker to becoming stronger than the expected limit, as $\Lambda$ grows, is a sign that we are soon entering the regime where the net effect of signal and destructive interference is a deficit.

For completeness, the limits obtained on $m_{q^*}$ from the angular distributions are also shown, despite not previously having been made public. This is a narrow resonance, with an expected exclusion limit at the final integrated luminosity at $m_{q^*} = 4.9$ TeV in the mass distribution analysis. The region in $m_{jj}$ populated by this signal lies within the $m_{jj}$ regions probed in the combined bin analysis. As Fig. 11.13 shows, the expected limit from the angular distributions is even slightly stronger, and the observed limit at $m_{q^*} = 5.2$ TeV coincides with the result from the mass distribution.

That the angular distribution is sensitive also to resonant phenomena may prove valuable in the case of an excess in the dijet mass distribution. If this is not seen in the angular distributions, it could mean that the process producing the excess is not more isotropic than QCD, which is either an important physics message or an indication of an experimental problem.

*DM Recast*

In addition to the CI limits, a DM recast of the limits on the CI $\Lambda$ can be done, following the method outlined in Ref. [1]. Here a maximum constraint on the coupling

---

[9]The combined fit increases the reach in $\Lambda$ by about 15% compared to the single-bin fit.

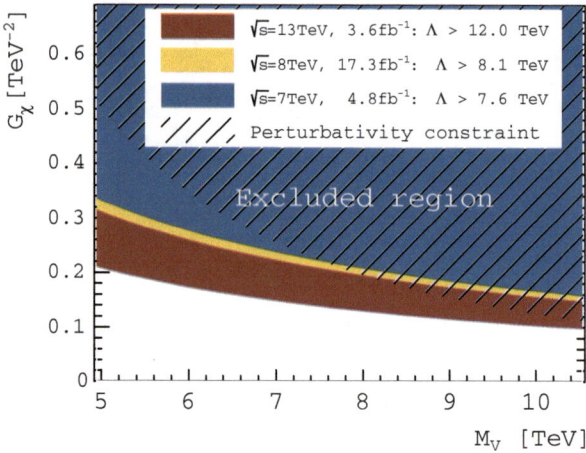

**Fig. 11.14** Upper limits on $G_\chi$ and $M_V$ derived from the limits on CI $\Lambda$ in the destructive interference scenario, obtained from angular distributions in $\sqrt{s} = 7, 8$ and 13 TeV data

strength between the mediator and both fermions and DM is set at $g \le \sqrt{4\pi}$, from the assumption that the interaction is perturbative. It is shown that, by identifying the effective CI coupling as

$$G_q \le \frac{4\pi}{\Lambda^2} \tag{11.2}$$

one can obtain 2-dimensional limits on the DM effective coupling $G_\chi$ and mediator mass $M_V$ through

$$G_\chi \le \frac{1}{M_V} \frac{4\pi}{\Lambda}. \tag{11.3}$$

The resulting exclusion limits from the limits on $\Lambda$ in a destructive interference scenario are shown here for the first time, for three[10] ATLAS results in Fig. 11.14. The horizontal axis is truncated at 5 TeV to ensure that the EFT approach is valid also at $\sqrt{s} = 13$ TeV. The destructive interference limits are used here as they are more conservative than the constructive interference ones.

## 11.3  Discussion

We have seen that the shape of the angular distributions shows a remarkable independence of $m_{jj}$. Given the impact on the shape from the removal of the masked modules in the $\sqrt{s} = 8$ TeV data, which lowers the low-$\chi$ region, the $\sqrt{s} = 8$ and 13 TeV angular distributions are not in agreement. However, comparing to the data dis-

---

[10]The results are obtained using the full 2011 [2], 2012 and 2015 data sets.

tributions obtained at $\sqrt{s} = 7$ TeV (see Ref. [2]),[11] they agree very well (within uncertainties) across $m_{jj}$, as well as with the $\sqrt{s} = 13$ TeV result. We thus see independence of both $m_{jj}$ and $\sqrt{s}$. In fact, this relates to several points that we have touched upon previously. One is the small PDF uncertainty in the angular distributions: which part of the PDF we are probing, be it with a different $\sqrt{s}$ or different PDF distribution, does not affect the shape of the angular distribution. Following the discussion of factorisation and the argument made in Sect. 8.2, this either implies a miraculous cancellation of hard-scatter and non-perturbative effects, or, it means that the non-perturbative effects are negligible and the partonic cross section is the same: no new scales are entering, and we are probing the same matrix elements. Put differently, we see scaling in the angular distributions. In particular, no new structure occurs as we probe smaller scales than ever before. Another implication of this relates to $F_\chi$. Given its definition, it is the ratio of the integral of the first four $\chi$ bins of the angular distribution, to the total integral. Since the shape of the angular distribution is independent of $m_{jj}$, $F_\chi$ will correspondingly be flat in $m_{jj}$. This is also what was seen in the $\sqrt{s} = 8$ TeV analysis (cf. Fig. 11.3).

That there is no deviation from the SM prediction in the angular distributions, could mean that the new phenomena they are sensitive to do not exist at the energy scale these distributions probe, or not with large enough cross section, or not at all. The large leap in sensitivity from the increased $\sqrt{s}$ between 2012 and 2015 enabled extending the reach in parameter space for many models with comparatively little data, but from here on, progress with the dijet final state will be slower. Apart from slowly reducing the statistical uncertainty, one can consider enhancing the sensitivity to new phenomena by retaining more information about the final state, or by employing a methodology that reduces the systematic uncertainties. The latter point will be discussed shortly. The former point relates to using single-jet or event-level observables in addition to the dijet observables of mass and angular separation explored thus far. One idea is to also measure the single-jet mass for high $m_{jj}$, low $\chi$, and compare to signal-free regions. This procedure requires understanding the compatibility across phase space regions, which can be explored in MC. Furthermore, it requires a good jet mass calibration and introduces the corresponding systematic uncertainty.

### 11.3.1  Outlook for Methodology Improvements

The independence of $m_{jj}$ strongly suggests that a data-driven SM estimate could be used, based on comparing the shapes across different $m_{jj}$—or even $\sqrt{s}$—to find the most discrepant region. In the version of $F_\chi$, a simple implementation is a low-order polynomial fit; for instance a constant, or a first-order polynomial allowing for a

---

[11]In hindsight, the agreement between data and MC would likely have improved already at $\sqrt{s} = 7$ TeV if the EW corrections had been available at the time.

small slope in the case of modulation from detector effects.[12] Here the reasoning follows that of the dijet mass spectrum fit in the search for resonant deviations. Given the more complicated shape of a falling spectrum, it uses a higher-order parameterisation, assuming that detector effects don't introduce bumps. For $F_\chi$, with its simpler shape, "any" smooth fit would work. However, maintaining sensitivity to non-resonant effects requires some rigidity compared to the dijet mass spectrum fit.

For the angular distributions in $\chi$, some tests have been performed calculating the $\chi^2$ probability for each $m_{jj}$ region of a prediction based on the others, as drawn in the distributions in Figs. 11.2 and 11.6. However, this is a global assessment of the agreement over all $\chi$, and better sensitivity could likely be obtained if only the low-$\chi$ region is allowed to contribute—agreement at high $\chi$ is guaranteed by construction!

If the MC prediction is retained and one were to pursue an analysis of $F_\chi$, the statistical modelling problem needs to be addressed. The ratio between the $m_{jj}$ distributions for two regions of $y^*$ follows a binomial distribution around a small number ($F_\chi \approx 0.07$). However, the statistical framework expects integer observables that are Poisson distributed, and this is the model constructed for the angular distributions (using $N$ vs $\chi$ with distributions normalised to have the same integral in all predictions as in data). One could construct analogous "$F_\chi$-like" distributions, using only two bins in $\chi$, corresponding to the first four current bins, and the rest.[13] This would again allow for a finer binning in $m_{jj}$; the one derived from the dijet mass resolution would be a reasonable choice. Deriving corrections and systematic uncertainties in such fine binning is somewhat computationally costly, however, all uncertainties not dominated by statistical uncertainties show smooth evolution in $m_{jj}$, making interpolations feasible.

A data-driven SM prediction would make the angular distribution search limited by statistical rather than systematic uncertainties. Given the overwhelming jet production at a hadron collider, and the outlook of much larger LHC data sets to come, the sensitivity to deviations in the lower $m_{jj}$ regions would be fantastic. This also means that all detector effects have to be very well understood, as small deviations would rapidly become highly significant. We have seen that the major concern would be a gradual miscalibration at high $p_T$, since this would cause migrations across $m_{jj}$ regions, affecting the $\chi$ shape. We have also seen that the large uncertainty on this effect as estimated in the JES uncertainty, is not supported in data, given the $m_{jj}$ independence of the shape, and the constraint from profiling across $m_{jj}$ on precisely this uncertainty. It thus seems like the ATLAS calorimeters and jet calibration are

---

[12]Given the successful prediction of scale invariance in the fully corrected MC prediction, albeit at a slightly different shape, assessing the impact of detector effects in MC is entirely feasible. It could be used either for correction of irregular effects, or for producing angular distributions that are fully corrected to particle-level (so-called *unfolding* of detector effects). One should note that unfolding always introduces additional uncertainties, making it preferable to search for new phenomena on detector-level distributions, retaining their statistical power.

[13]The reason why $F_\chi$ uses the inclusive $y^*$ in the denominator is to avoid division by zero in the case of signal-like entries only. This is not a concern in the setup proposed here and thus the "exclusive" outer region $0.6 < |y^*| < 1.7$ (or corresponding optimised choices) should be used to preserve as much shape information as possible.

both performing very well, giving enough confidence to encourage the exploration of data-driven methods in the data sets to come.

## References

1. H. Dreiner, D. Schmeier, J. Tattersall, Contact interactions probe effective dark matter models at the LHC. Europhys. Lett. **102**, 51001 (2013)
2. ATLAS Collaboration, ATLAS search for new phenomena in dijet mass and angular distributions using pp collisions at $\sqrt{s} = 7$ TeV. JHEP **1301**, 029 (2013)

# Chapter 12
# Conclusions and Outlook

We are still in the early days of LHC operation. Since 2010, this accelerator has seen two collision energy increases: from 7 to 8 in 2012, and then to 13 TeV in 2015. This thesis describes one of the best early measurements to do when the collision energy increases: dijet searches. Best because it reaches far into new energy domains, benefits greatly from increased centre-of-mass energies, and uses the most abundant final state at a hadron collider. It can thus break new ground already with a very small data set.

The data at $\sqrt{s} = 8$ and 13 TeV were used to search for phenomena beyond the Standard Model, exploiting the fact that the QCD prediction is well known: the dijet mass spectrum is featureless, and the dijet angular distribution in QCD is dominated by small scattering angles. The search looks for an onset of more isotropic dijet production at some new scale in physics. Even if the searches described here probed new phase space, the shape of the angular distributions was seen to be largely independent of dijet masses and centre-of-mass energies. This means that even at the smallest scales probed to date, the proton constituents show no new structure.

At both energies, the distributions were well described by the SM prediction, both in terms of a smooth fit to the dijet mass spectrum and a MC prediction for the angular distributions, which are the focus of this thesis. The results were thus used to set new limits on parameters of models of phenomena beyond the Standard Model. Models for quark compositeness and strong gravity, both types introduced as solutions to the hierarchy problem, are used as benchmark models. The resulting limits on the quark compositeness scale $\Lambda$ in a Contact Interaction model obtained in the 2015 data set at $\sqrt{s} = 13$ TeV are the strongest to date. A combined fit across $m_{jj}$ regions improved the reach in $\Lambda$ compared to the single-bin method used in the $\sqrt{s} = 8$ TeV data. In particular an enhanced sensitivity to resonant phenomena was achieved. Limits on the masses of both quantum black hole in an $n = 6$ ADD scenario and excited quarks were thus set with the dijet angular distributions. In addition, a dark matter interpretation of the limits obtained on the Contact Interaction $\Lambda$ were shown here for the first time.

© Springer International Publishing AG 2017                                                          159
L.K. Bryngemark, *Search for New Phenomena in Dijet Angular Distributions
at √s = 8 and 13 TeV*, Springer Theses, DOI 10.1007/978-3-319-67346-2_12

For the first time, EW corrections were included in the SM prediction of the angular distributions. These corrections are substantial and bring the MC prediction to the same $m_{jj}$ independence of angular distributions as observed in data. With the new results in hand, this $m_{jj}$ independence is now seen across several $\sqrt{s}$ and $m_{jj}$ ranges, and it is suggested to exploit this for a data-driven SM prediction in coming dijet angular distribution searches.

# Appendix A
# Simulation Settings

## A.1 QCD Prediction

### *A.1.1 8 TeV*

PYTHIA8 with AU2 [1] underlying event tune and leading-order CT10 [2] PDFs.

### *A.1.2 13 TeV*

PYTHIA8 with A14 [3] underlying event tune and leading-order NNPDF2.3 [4, 5] PDFs.

## A.2 Signal Simulation

The CI and $q^*$ simulation use the same choice for tune and PDF as the corresponding PYTHIA8 prediction. All samples are fully simulated using GEANT4.

For QBH, the CTEQ6L1 [6, 7] PDFs are used, and it was verified that the results from so-called fast detector simulation were equivalent to those obtained using the full GEANT4 simulation. Thus for speed, some mass points use fast detector simulation.

© Springer International Publishing AG 2017
L.K. Bryngemark, *Search for New Phenomena in Dijet Angular Distributions at √s = 8 and 13 TeV*, Springer Theses, DOI 10.1007/978-3-319-67346-2

## A.3 *K*-factor Calculations

### A.3.1 PYTHIA8 *LO* + *Parton Shower Settings*

All settings equal to the nominal PYTHIA8 prediction, but with these settings to make sure the processing is halted after parton showers:

```
pythia.readString("HardQCD:all = on");
pythia.readString("PartonLevel:FSR = on");
pythia.readString("PartonLevel:ISR = on");
pythia.readString("PartonLevel:MPI = off");
pythia.readString("SpaceShower:QCDshower = on");
pythia.readString("BeamRemnants:primordialKT = off");
pythia.readString("HadronLevel:all = off");
```

### A.3.2 NLOJET++

#### $\sqrt{s} = 8$ TeV

The settings for the NLOJET++ calculations are: PDF set 0 from CT10, $\alpha_s(M_Z) = 0.118$ with NLO precision in the running of $\alpha_s$, and the renormalisation and factorisation scales $\mu_R$ and $\mu_F$ are both set to $p_T^{avg} = (p_T^{lead} + p_T^{sublead})/2$.

#### $\sqrt{s} = 13$ TeV

Like for $\sqrt{s} = 8$ TeV running, but with PDFs from NNPDF2.3, with NLO precision in the running of $\alpha_s$.

## A.4 PDF Uncertainty Calculation

The PDF uncertainty is calculated using NLOJET++ connected to APPLgrid. Three PDFs are considered: CT10, MSTW2008 [8], and NNPDF2.3. The error members of each PDF are used to calculate that PDF's uncertainty through the envelope method [9] in Eq. A.1, where $X[q^{(i)}]$ is the cross section evaluated at the member set $i$.

$$\sigma^{CT10,\ MSTW}(PDF, \pm) =$$

$$f \cdot \sqrt{\sum_{i=1}^{N/2} (\max(\{\pm X[\{q^{(2i-1)}\}] \mp X[\{q^{(0)}\}]), (\pm X[\{q^{(2i)}\}] \mp X[\{q^{(0)}\}]), 0\})^2}$$

$$\sigma^{NNPDF}(PDF, \pm) = \sqrt{\frac{1}{N-1}\sum_{i=1}^{N} X[\{q^{(i)}\}] - X[\{q^{(0)}\}]}, \qquad (A.1)$$

where $N$ corresponds to different sizes of member sets for the different PDF sets. For CT10, a rescaling factor of $f = \frac{1}{1.64485}$ is included to scale the CT10 uncertainty from 90 to 68% C.L., while MSTW PDF has no need for rescaling ($f = 1$). The uncertainties of all three PDFs are then combined through Eqs. A.2–A.4, where $j$ is the PDF set ($j$ = CT10, MSTW, NNPDF). The final uncertainties are given by $\delta(\pm)$.

$$U = \max\{X^j[\{q^{(0)}\}] + \sigma^j(PDF, +)\} \qquad (A.2)$$

$$L = \min\{X^j[\{q^{(0)}\}] - \sigma^j(PDF, -)\} \qquad (A.3)$$

$$M = \frac{U+L}{2}; \quad \delta(+) = \frac{U-M}{M}; \quad \delta(-) = \frac{L-M}{M} \qquad (A.4)$$

# References

1. ATLAS Collaboration. Summary of ATLAS Pythia 8 tunes. Tech. rep. ATL-PHYS-PUB-2012-003. Geneva: CERN, 2012.
2. Hung-Liang Lai et al. "New parton distributions for collider physics". In: Phys.Rev. D82 (2010), p. 074024.
3. ATLAS Collaboration. ATLAS Run 1 Pythia8 tunes. Tech. rep. ATL-PHYS-PUB-2014-021. Geneva: CERN, 2014.
4. Christopher S. Deans. "Progress in the NNPDF global analysis". In: Proceedings, 48th Rencontres de Moriond on QCD and High Energy Interactions (2013), pp. 353–356.
5. Stefano Carrazza, Stefano Forte, and Juan Rojo. "Parton Distributions and Event Generators". In: Proceedings, 43rd International Symposium on Multiparticle Dynamics (ISMD 13). 2013, pp. 89–096.
6. S. Kretzer et al. "CTEQ6 parton distributions with heavy quark mass effects". In: Phys. Rev. D69 (2004), p. 114005.
7. J. Pumplin et al. "New generation of parton distributions with uncertainties from global QCD analysis". In: JHEP 07 (2002), p. 012.
8. A. D. Martin et al. "Parton distributions for the LHC". In: Eur. Phys. J. C63 (2009), pp. 189–285.
9. Michiel Botje et al. "The PDF4LHC Working Group Interim Recommendations". arXiv:1101.0538. 2011.

# Appendix B
# Data Set and Event Selection Details

This section gives the technical details for data set selection for the two searches.

## B.1 8 TeV

The data used in the 8 TeV analysis correspond to a total integrated luminosity of $17.3\,\mathrm{fb}^{-1}$. The data samples employed are the centrally produced NTUP-SLIMSMQCD slims for the JetTauETMiss and Had Delayed streams, together with the full NTUP-COMMON for the debug stream. The event selection (below) is based on the following GRL:

```
data12_8TeV.periodAllYear_DetStatus-v61-pro14-
02_DQDefects-00-01-00_PHYS_StandardGRL_All_Good.xml.
```

Jet calibration uses the ATLAS recommendation for Moriond 2013: tag `00-08-15` for uncertainties and

```
ApplyJetCalibration-00-03-03/
JES_Full2012dataset_Preliminary_Jan13.config
```

for calibration.[1]

### B.1.1 Analysis Cutflow

The analysis selection criteria below are applied to collision data events in the listed order. These criteria are repeated in Table B.1, which tabulates the number of events surviving each cut, $N_{ev}$. This section states the final numbers after removing the blinding cuts used in the optimisation phase of the analysis.

---

[1] See the ATLAS internal
https://twiki.cern.ch/twiki/bin/viewauth/AtlasProtected/JetUncertainties2012 for further information.

© Springer International Publishing AG 2017
L.K. Bryngemark, *Search for New Phenomena in Dijet Angular Distributions
at √s = 8 and 13 TeV*, Springer Theses, DOI 10.1007/978-3-319-67346-2

1. Total events in the data sample, $N_{ev}$, using the NTUP_SLIMSMQCD data format.
2. Jets are recalibrated at this point. If the data sample is to be reduced by blinding (eg., every 4th event), it is done here.
3. Events must pass the trigger requirements. The trigger uses 11 single-jet triggers, covering 11 contiguous, non-overlapping $p_T$ ranges with an efficiency of 99.5% or greater. (The trigger strategy is described in Sect. 10.1.2.)
4. Events must be from runs appearing in the GRL named above.
5. Require the first vertex to have $N_{track} > 1$.
6. Reject events with calorimeter data integrity problems: larError = 2, OR tileError = 2. Also reject incomplete events (where some detector information is missing), by checking the CoreFlags that would indicate this condition.
7. Reject events if there has been a calorimeter module trip, as indicted by TTileTripReader.
8. Require two leading jets, both within the range $|y| < 2.8$.
9. Reject the event if either the leading or subleading jet is associated with a Tile calorimeter hotspot.[2]
10. Reject events where either of the two leading jets is ugly, or if any other jet is ugly and has
    $p_T > 0.3 \times p_T^{sublead}$.[3]
11. Reject the event if either the leading or subleading jet is a bad jet, as determined by the Rel 17 BadLooser definition.[4]
12. Reject events where either of the two leading jets, or any other jets with $p_T > 0.3 \times p_T^{sublead}$, falls within a Tile module that is masked, as stated in Sect. 10.2.2.
13. Require each of the leading jets to have $p_T > 50$ GeV. This selection criterion, and those that follow, are applied to jets that have been corrected for the pile-up energy, and calibrated to the hadronic scale.
14. Beginning here, selection criteria are applied to dijet variables. Retain events with $|y^*| = |y_1 - y_2|/2 < 1.7$.
15. Retain events with $|y_B| = |y_1 + y_2|/2 < 1.1$.
16. Retain events with $m_{jj} > 600$ GeV.

The cut flow for the full data sample used is shown in Tables B.1 and B.2, for the overlap and delayed stream, respectively. This doesn't include the additional 24 events from the debug stream, which are also used. The debug stream contains events, in which the trigger was not able to make a decision online.[5] Those are reprocessed later. If those events pass the trigger decision in the reprocessing, they are stored in debugrec_hltacc data. The 24 events quoted are the events passing event

---

[2]More information at: https://twiki.cern.ch/twiki/bin/view/AtlasProtected/ExoticDijets2012Cut flows\#Tile_hotspot_cleaning_for_Period.

[3]More information at: https://twiki.cern.ch/twiki/bin/viewauth/AtlasProtected/HowToCleanJets 2012.

[4]More information at: https://twiki.cern.ch/twiki/bin/viewauth/AtlasProtected/HowToCleanJets 2012.

[5]See https://twiki.cern.ch/twiki/bin/view/Atlas/DebugStream.

**Table B.1** Cut flow for the full data sample used (overlap stream), showing $N_{ev}$ and the cut efficiency for every cut in the analysis

| Selection criterion | $N_{ev}$ | Rel. change[%] |
| --- | --- | --- |
| 1 (before cuts) | 796605696 | 0.00 |
| 2 (blinding, if enabled) | 796605696 | 0.00 |
| 3 (trigger check) | 2390060 | −99.70 |
| 4 (after GRL) | 2029261 | −15.10 |
| 5 (vertex check) | 2029243 | $-8.87 \cdot 10^{-4}$ |
| 6 (calorimeter error cut) | 2025437 | −0.19 |
| 7 (TileTripReader cut) | 2025437 | −0.00 |
| 8 (2 leading jets and y cut) | 1769276 | −12.65 |
| 9 (Tile hotspot check) | 1769276 | 0.00 |
| 10 (after ugly jet cut) | 1745377 | −1.35 |
| 11 (after bad jet cut) | 1744395 | −0.06 |
| 12 (masked Tile module cut) | 1542609 | −11.57 |
| 13 (after jet $p_T$ cut) | 1120754 | −27.35 |
| 14 (after y*cut) | 1082194 | −3.44 |
| 15 (after $y_B$ cut) | 759379 | −29.83 |
| 16 (after $m_{jj}$ cut) | 209698 | −72.39 |

**Table B.2** Cut flow for the full data sample used (delayed stream), showing $N_{ev}$ and the cut efficiency for every cut in the analysis

| Selection criterion | $N_{ev}$ | Rel. change[%] |
| --- | --- | --- |
| 1 (before cuts) | 417493056 | 0.00 |
| 2 (blinding, if enabled) | 417493056 | 0.00 |
| 3 (trigger check) | 33883520 | −91.88 |
| 4 (after GRL) | 32385738 | −4.42 |
| 5 (vertex check) | 32385524 | $-6.61 \cdot 10^{-4}$ |
| 6 (calorimeter error cut) | 32285032 | −0.31 |
| 7 (TileTripReader cut) | 32285008 | $-7.43 \cdot 10^{-5}$ |
| 8 (two leading jets and y cut) | 32225564 | −0.18 |
| 9 (Tile hotspot check) | 32225564 | 0.00 |
| 10 (after ugly jet cut) | 32221578 | −0.01 |
| 11 (after bad jet cut) | 32201909 | −0.06 |
| 12 (masked Tile module cut) | 28513018 | −11.46 |
| 13 (after jet $p_T$ cut) | 28466487 | −0.16 |
| 14 (after y* cut) | 28108922 | −1.26 |
| 15 (after $y_B$ cut) | 23651795 | −15.86 |
| 16 (after $m_{jj}$ cut) | 22090360 | −6.60 |

selection. The events from this stream include one event at 3.9 TeV, while the other 23 events are below 2.6 TeV.

Since the final distributions are all normalised to unit area, it may be interesting to see the how the actual number of events after full selection are distributed among the $m_{jj}$ bins used in the analysis. These numbers are given in Table B.3.

For collision data samples, all selection criteria are applied. For MC samples (signal and QCD) we only apply the kinematic selection, and the emulation of the masked Tile calorimeter regions as explained in Sect. 10.2.2.

## B.2 13 TeV

- Good Run List (GRL): Requirement that all relevant detectors were in a good state ready for physics
- LAr: Liquid Argon Calorimeter error rejected
  ( errorState(xAOD::EventInfo::LAr) )
- Tile: Tile Calorimeter error rejected
  ( errorState(xAOD::EventInfo::Tile) )
- Core: Incomplete event build rejected
  ( isEventFlagBitSet(xAOD::EventInfo::Core, 18) )
- Primary Vertex: the highest $\sum p_T^2(trk)$ vertex has at least two tracks associated with it
  (xAOD::VxType::VertexType::PriVtx)
- Trigger: passes OR of L1_J75, L1_J100, HLT_J360, HLT_J380, HLT_J400
- at least two clean jets with $p_T > 50$ GeV
- Leading jet $p_T > 440$ GeV
- $|y^*| < 1.7$
- $|y_B| < 1.1$
- $m_{jj} > 2500$ GeV

**Table B.3** Final $N_{ev}$, for the full data sample used, in each $m_{jj}$ interval used for binning the angular distributions

| $m_{jj}$ range [GeV] | $N_{ev}$ |
|---|---|
| 600–800 | 8571722 |
| 800–1200 | 9917319 |
| 1200–1600 | 2825705 |
| 1600–2000 | 756044 |
| 2000–2600 | 183829 |
| 2600–3200 | 19609 |
| 3200–8000 | 2550 |

The GRL xml file is
`data15_13TeV.periodAllYear_DetStatus-v65-pro19-`
`01_DQDefects-00-01-02_PHYS_StandardGRL_All_Good.xml`
The information in parenthesis is technical information related to the xAOD EDM.

## B.2.1 Analysis Selection and Cutflow

The data selection cutflow is shown in Table B.4 and the MC selection cutflow in Table B.5. The number of events passing the selection in MC is rounded to the nearest integer (the effective statistics vary between slices).

**Table B.4** Cutflow for data events with the analysis cuts used for $\sqrt{s} = 13$ TeV data. "Trigger" corresponds to the events passing the OR of L1_J75, L1_J100, HLT_J360, HLT_J380, HLT_J400

| Selection criteria | $N_{events}$ | Rel. decrease (%) |
|---|---|---|
| all | 35477718 | 0.0 |
| LAr | 35398888 | −0.22 |
| tile | 35395678 | −0.01 |
| core | 35393381 | −0.01 |
| NPV | 35391453 | −0.01 |
| Trigger (OR) | 23350594 | −34.02 |
| jetSelect signal | 23020926 | −1.41 |
| $jet_1 p_T > 200$ GeV | 12740838 | −44.66 |
| HLT j360 | 11995952 | −5.85 |
| cleaning | 11988448 | −0.06 |
| LJetPt | 4979860 | −58.46 |
| mjjMin | 136300 | −97.26 |
| y* < 1.7 | 71204 | −47.76 |
| yBoost < 1.1 | 70417 | −1.11 |

**Table B.5** Cutflow for PYTHIA8 events in the analysis of $\sqrt{s} = 13$ TeV data

| Selection criteria | $N_{events}$ | Rel. decrease (%) |
|---|---|---|
| HLT j360 | 4804917 | 0.0 |
| LJetPt | 1884597 | −60.78 |
| mjjMin | 61818 | −96.72 |
| y* < 1.7 | 29333 | −52.55 |
| yBoost < 1.1 | 28880 | −1.54 |

# Appendix C
# LHCP Results

Here the results prepared for the 2015 LHCP conference are shown (Figs. C.1 and C.2). These were the first search results to be approved by ATLAS in Run2 and represent 80 pb$^{-1}$ of data at $\sqrt{s} = 13$ TeV [1]. No significant deviations from the SM prediction were found, and limits were set on QBH as modelled by BlackMax and QBH, which surpassed the limits obtained in Run1 using 20.2 fb$^{-1}$ at $\sqrt{s} = 8$ TeV. All details can be found in Ref. [1]. The only difference in methodology to the full 2015 data set result is the search strategy in the angular distributions: this result uses a single region $m_{jj} > 3.4$ TeV for the statistical analysis, while the analysis of the full 2015 data set uses a combination of the four highest regions in $m_{jj}$, starting from $m_{jj} = 3.4$ TeV (Figs. C.1 and 11.5).

© Springer International Publishing AG 2017
L.K. Bryngemark, *Search for New Phenomena in Dijet Angular Distributions
at $\sqrt{s} = 8$ and 13 TeV*, Springer Theses, DOI 10.1007/978-3-319-67346-2

**Fig. C.1** Normalised angular distributions in the $\sqrt{s} = 13$ TeV data, overlaid with the MC prediction. Theoretical and total uncertainties are shown as lighter and darker shaded bands, while the vertical error bars represent the statistical uncertainty. The predicted signal for QBH is also shown

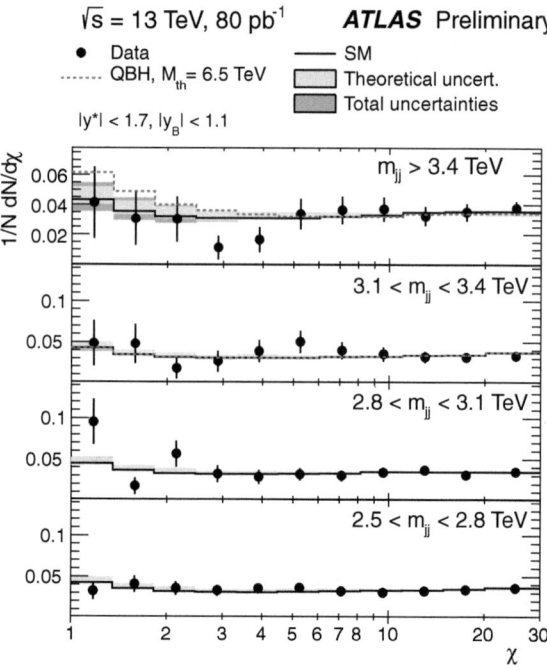

**Fig. C.2** Dijet mass distribution in the $\sqrt{s} = 13$ TeV data, compared to the fit (red line) in the second panel, with QBH signal prediction for two $M_{th}$ overlaid (color figure online)

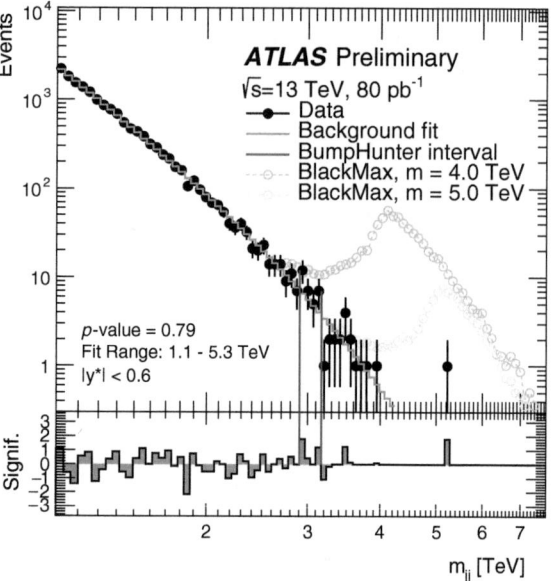

# Reference

1. ATLAS Collaboration, Search for New Phenomena in Dijet Mass and Angular Distributions with the ATLAS Detector at $\sqrt{s} = 13$ TeV. Tech. rep. ATLASCONF-2015-042. Geneva: CERN, 2015.